Practical Quantum Computing for Developers

Programming Quantum Rigs in the Cloud using Python, Quantum Assembly Language and IBM QExperience

Vladimir Silva

Apress®

Practical Quantum Computing for Developers: Programming Quantum Rigs in the Cloud using Python, Quantum Assembly Language and IBM QExperience

Vladimir Silva
CARY, NC, USA

ISBN-13 (pbk): 978-1-4842-4217-9   ISBN-13 (electronic): 978-1-4842-4218-6
https://doi.org/10.1007/978-1-4842-4218-6

Library of Congress Control Number: 2018966346

Managing Director, Apress Media LLC: Welmoed Spahr
Acquisitions Editor: Steve Anglin
Development Editor: Matthew Moodie
Coordinating Editor: Mark Powers

Cover designed by eStudioCalamar

Cover image designed by Freepik (www.freepik.com)

Distributed to the book trade worldwide by Springer Science+Business Media New York, 233 Spring Street, 6th Floor, New York, NY 10013. Phone 1-800-SPRINGER, fax (201) 348-4505, e-mail orders-ny@springer-sbm.com, or visit www.springeronline.com. Apress Media, LLC is a California LLC and the sole member (owner) is Springer Science + Business Media Finance Inc (SSBM Finance Inc). SSBM Finance Inc is a **Delaware** corporation.

For information on translations, please e-mail editorial@apress.com; for reprint, paperback, or audio rights, please email bookpermissions@springernature.com.

Apress titles may be purchased in bulk for academic, corporate, or promotional use. eBook versions and licenses are also available for most titles. For more information, reference our Print and eBook Bulk Sales web page at http://www.apress.com/bulk-sales.

Any source code or other supplementary material referenced by the author in this book is available to readers on GitHub via the book's product page, located at www.apress.com/9781484242179. For more detailed information, please visit http://www.apress.com/source-code.

Table of Contents

About the Author

Vladimir Silva holds a Master's degree in Computer science from Middle TN State University. He worked for 5 years for IBM as a Research Engineer where he acquired extensive experience in distributed and Grid computing.

He holds numerous IT certifications, including OCP, MCSD, and MCP, and has written many technical articles for IBM developerWorks. His previous books include *Grid Computing for Developers* (Charles River Media), *Practical Eclipse Rich Client Platform* (Apress), *Pro Android Games* (Apress), and *Advanced Android 4 Games* (Apress).

An avid marathon runner, with over 16 races completed all over the state of NC (by the time of this writing), when not coding, writing or running he enjoys playing his classic guitar and pondering about awesome things like Quantum Mechanics.

About the Technical Reviewer

Jason Whitehorn is an experienced entrepreneur and software developer and has helped many oil and gas companies automate and enhance their oilfield solutions through field data capture, SCADA, and machine learning. Jason obtained his Bachelor of Science in Computer Science from Arkansas State University, but he traces his passion for development back many years before then, having first taught himself to program BASIC on his family's computer while still in middle school.

When he's not mentoring and helping his team at work, writing, or pursuing one of his many side projects, Jason enjoys spending time with his wife and four children and living in the Tulsa, Oklahoma region. More information about Jason can be found on his website: https://jason.whitehorn.us

Introduction

I wrote this book to be the ultimate guide for programming a quantum computer in the cloud. Thanks to the good folks at IBM Research, this is now possible. IBM has made their prototype quantum rig (known as the IBM Q Experience) available not only for research but for individuals in general interested in this field of computing.

Quantum computing is gaining traction, and now is the time to learn to program these machines. In years to come, the first commercial quantum computers should be available, and they promise significant computational speedups compared to classical computers. Consider the following graph showing the time complexities for two large integer factorization algorithms: the best classical algorithm, the Number Field Sieve, vs. the quantum factorization algorithm developed by Peter Shor.

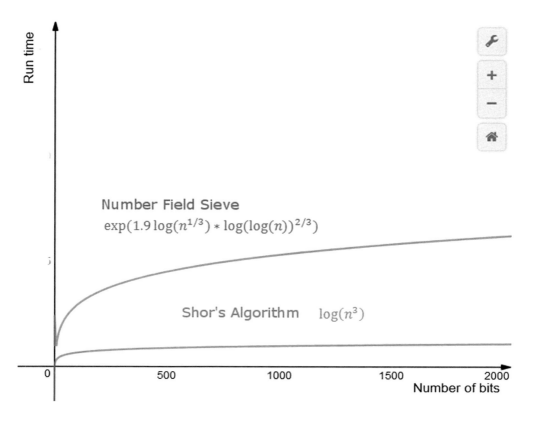

Shor's algorithm provides a significant speedup over the Number Field Sieve for a problem, that is, the foundation of current cryptography. A practical implementation of this algorithm will render current asymmetric encryption useless!

All in all, this book is a journey of understanding. If you find the concepts explained throughout the chapters difficult to grasp, then you are not alone. The great physicist Richard Feynman once said: *If somebody tells you he understands quantum mechanics, it means he doesn't understand quantum mechanics.* Even the titans of this bizarre theory have struggled to understand what it all means.

I have tried to explore quantum computation to the best of my abilities by using real-world algorithms, circuits, code, and graphical results. Some of the algorithms included in this manuscript defy logic and seem more voodoo magic than a computational description of a physical system. This is the main reason I decided to tackle this subject. Even though I find it hard to understand the mind-bending principles of quantum mechanics, I've always been fascinated by this awesome theory. Thus when IBM came up with its one-of-a-kind quantum computing platform for the cloud, and opened it up for the rest of us, I jumped to the opportunity to learn and create this manuscript.

Ultimately, this is my take on quantum computing in the cloud, and I hope you find as much enjoyment reading it as I got writing it. My humble advice: Learn to program quantum computers; soon they will be ever present in the data center, doing everything from search and simulations to medicine and artificial intelligence. You name it. In general terms, the manuscript is divided into the following chapters:

Chapter 1: The Bizarre and Awesome World of Quantum Mechanics

It all began in the 1930s with Max Planck, the reluctant genius. He came up with a new interpretation for the energy distribution of the light spectrum. He started it all by unwillingly postulating that the energy of the photon was not described by a continuous function, as believed by classical physicists, but by tiny chucks he called *quanta*. He was about to start the greatest revolution in science in this century: *quantum mechanics*. This chapter is an appetizer to the main course and explores the clash of two titans of physics: Albert Einstein and Niels Bohr. Quantum mechanics was a revolutionary theory in the 1930s, and most of the scientific establishment was reluctant to accept it, including the colossus of the century: Albert Einstein. Fresh from winning the Nobel Prize, Einstein never accepted the probabilistic nature of quantum mechanics. This caused a rift with its biggest champion: Niels Bohr. The two greats debated it out for decades and never resolved their differences. Ultimately, quantum mechanics has withstood 70 years of theoretical and experimental challenges, to emerge always triumphant. Read this chapter and explore the theory, experiments, and results, all under the cover of the incredible story of these two extraordinary individuals.

Chapter 2: Quantum Computing: Bending the Fabric of Reality Itself

In the 1980s, another great physicist – Richard Feynman – proposes a quantum computer, that is, a computer that can take advantage of the principles of quantum mechanics to solve problems faster. The race is on to construct such a machine. This chapter explores, in general terms, the basic architecture of a quantum computer: qubits, the basic blocks of quantum computation. They may not seem like much, but they have almost magical properties: superposition; believe it or not, a qubit can be in two states at the same time: 0 and 1. This concept is hard to grasp at the macroscale where we live. Nevertheless, at the atomic scale, all bets are off. This fact has been proven experimentally for over 70 years. Thus superposition allows a quantum computer to outmuscle a classical computer by performing large amounts of computation with relatively small numbers of qubits. Another mind bender is qubit entanglement: something that, when explored, seems more like voodoo magic than a physical principle. Entangled qubits transfer states, when observed, faster than the speed of light across time or space! Wrap your head around that. All in all, this chapter explores all the physical components of a quantum computer: quantum gates, types of qubits such as superconducting loops, ion traps, topological braids, and more. Furthermore, the current efforts of all major technology players in the subject are described, as well as other types of quantum computation such as quantum annealing.

Chapter 3: Enter the IBM Q Experience: A One-of-a-Kind Platform for Quantum Computing in the Cloud

In this chapter, you will get your feet wet with the IBM Q Experience. This is the first quantum computing platform in the cloud that provides real or simulated quantum devices for the rest of us. Traditionally, a real quantum device will be available only for research purposes. Not anymore, thanks to the folks at IBM who have been building this stuff for decades and graciously decided to open it up for public use.

Learn how to create a quantum circuit using the visual Composer or write it down using the excellent Python SDK for the programmer within you. Then execute your circuit in the real thing, explore the results, and take the first step in your new career as a quantum programmer. IBM may have created the first quantum computing platform in the cloud, but its competitors are close behind. Expect to see new cloud platforms in the near future from other IT giants. Now is the time to learn.

Chapter 4: QISKit, Awesome SDK for Quantum Programming in Python

QISKit stands for Quantum Information Software Kit. It is a Python SDK to write quantum programs in the cloud or a local simulator. In this chapter, you will learn how to set up the Python SDK in your PC. Next, you will learn how the quantum gates are described using linear algebra to gain a deeper understanding of what goes on behind the scenes. This is the appetizer to your first quantum program, a very simple thing to familiarize yourself with the syntax of the Python SDK. Finally you will run it in a real quantum device. Of course, quantum programs can also be created visually in the Composer. Gain a deeper understanding of quantum gates, the basic building blocks of a quantum program. All this and more is covered in this chapter.

Chapter 5: Start Your Engines: From Quantum Random Numbers to Teleportation, Pit Stop at Super Dense Coding

This chapter is a journey through three remarkable information processing capabilities of quantum systems. Quantum random number generation explores the nature of quantum mechanics as a source for true randomness. You will learn how this can be achieved using very simple logic gates and the Python SDK. Next, this chapter explores two related information processing protocols: super dense coding and quantum teleportation. They have exuberant names and almost magical properties. Discover their secrets, write circuits for the Composer, execute remotely using Python, and finally interpret and verify their results.

Chapter 6: Fun with Quantum Games

In this chapter, you will learn how to implement a basic game in a quantum computer. For this purpose, we use the quintessential Quantum Battleship distributed with the QISKit Python tutorial. The first part looks at the mechanics of the game, yet we don't stop there. The second part of this chapter takes things to the next level by giving it a major face-lift. In this part, you will put Quantum Battleship in the cloud by giving it a browser-based user interface, an Apache CGI interface to consume events and dispatch them to the quantum simulator, and more. Impress your friends and family by playing Quantum Battleship with your web browsers in the cloud.

Chapter 7: Game Theory: With Quantum Mechanics, Odds Are Always in Your Favor

This is a weird one, even for quantum mechanics standards. This chapter explores two game puzzles that show the remarkable power of quantum algorithms over their classical counterparts: the counterfeit coin puzzle and the Mermin-Peres Magic Square. In the counterfeit coin puzzle, a quantum algorithm is used to reach quartic speedup over the classical solution for finding a fake coin using balance scale a limited number of times. The Mermin-Peres Magic Square is an example of quantum pseudo-telepathy or the ability of players to almost read each other's minds achieving outcomes only possible if they communicate during the game.

Chapter 8: Faster Search plus Threatening the Foundation of Asymmetric Cryptography with Grover and Shor

This chapter brings proceedings to a close with two algorithms that have generated excitement about the possibilities of practical quantum computation: Grover's search, an unstructured quantum search algorithm capable of finding inputs at an average of square root of N steps. This is much faster than the best classical solution at N/2 steps. It may not seem that much, but, when talking about very large databases, this algorithm can crush it in the data center. Expect all web searches to be performed by Grover's in the future. Shor's integer factorization: The notorious quantum factorization that experts say could bring current asymmetric cryptography to its knees. This is the best example of the power of quantum computation by providing exponential speedups over the best classical solution.

## CHAPTER 1

The Bizarre and Awesome World of Quantum Mechanics

The story of quantum mechanics is a fable of wonder and bewilderment. It has elements of science, philosophy, religion, and dare I say magic. It'll turn your mind upside down, and sometimes it'll make you question the existence of an all-powerful creator out there. Even though I find its concepts difficult to grasp, I've always been fascinated by it. Some of the concepts presented in this chapter are hard to understand; however don't be troubled. Nobody has been able to fully describe what this all means, not even the titans of physics fully understand quantum mechanics. However that doesn't mean we can't be fascinated by it. The great physicist Richard Feynman once said: If somebody tells you he understands quantum mechanics, it means he doesn't understand quantum mechanics. This chapter is my take on this fascinating fable and how the struggle of two titans of science shaped its past, present, and future.

It all began in the 1930s, after Albert Einstein rose to world fame with the theory of special relativity which built upon Newtonian physics to unify the heavens and the earth. While Einstein was looking to the heavens, a new breed of scientists were looking at the very small. Spearheaded by giants of physics such as Max Planck, Ernest Rutherford, and Niels Bohr, it started a clash of titans and one of the greatest debates of physics in the twentieth century – on one side, Albert Einstein, fresh from winning the Nobel Prize for his groundbreaking discoveries on the nature of light and special relativity and, on the other side, Niels Bohr, whose contributions to the field of quantum mechanics would earn him a Nobel Prize in 1922 and the prestigious *Order of the Elephant*, a Danish distinction normally reserved for royalty. Let's take a look how the struggle between these two greats shaped the science masterpiece, that is, quantum mechanics.

© Vladimir Silva 2018
V. Silva, *Practical Quantum Computing for Developers*, https://doi.org/10.1007/978-1-4842-4218-6_1

The Golden Age of Physics in the Twentieth Century

At the beginning of the twentieth century, British scientist Ernest Rutherford made a startling discovery about the nature of the atom. He postulated that atoms look like tiny solar systems, made of a tiny nucleus with positive charge and electrons negatively charged rotating like tiny planes around it. This was a remarkable insight as it was previously believed that the atom was a simple spherical blob of mass with positive and negative charges.

Bohr arrived at Rutherford's lab in Cambridge in 1920 and fell in love with Rutherford's model of the atom, but there was a problem, and a big one. If classical Newtonian physics are applied to Rutherford's model where negatively charged electrons rotate around a positively charged nucleus, the electron will eventually fall inside and crash against the nucleus creating a catastrophic paradox. Nothing should exist, as electrons will crash in a matter of seconds. Bohr saw this, and with undeterred excitement, he delayed his marriage and canceled his honeymoon in an effort to save Rutherford's model. Bohr postulates in a paper that electrons move in fixed orbits that cannot change. This goes against the basis of Newtonian physics but draws upon new ideas from the father of quantum mechanics, Max Planck.

Max Planck and the Ultraviolet Catastrophe Started It All

Planck suggested that heat and light come in units that cannot be divided, which he called "energy quanta." Planck's idea came from his efforts to solve the black-body radiation experiments where a body that completely absorbs all radiation (heat) inside it has a cavity that allows some radiation to escape (see Figure 1-1). As the heat increases inside the box, the frequency of the radiation reaches ranges visible to the human eye, glowing at different colors. It was well known by porcelain makers at the time that all bodies produce fixed colors at given temperatures (see Table 1-1).

Black Body Radiation

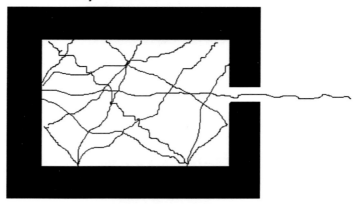

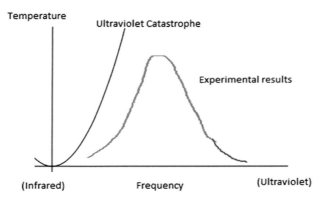

Figure 1-1. *Black-body radiation experiment results*

Table 1-1. *Light Colorization at Different Temperatures*

Temperature (°C)	Color
500	Dark red
800	Cherry red
900	Orange
1000	Yellow
1200	White

Figure 1-1 shows the black-body radiation experiment along with the results provided by the classical theory of radiation curves collected from experiments in the 1890s. Classical physicist's experiments predicted infinite intensities for the ultraviolet spectrum. This became known as the ultraviolet catastrophe and was the product of dubious theoretical arguments and experimental results. If true, this would mean, for example, that it will be dangerous to seat anywhere close to a fireplace! Planck sought to find a solution to the ultraviolet catastrophe.

Planck used the second law of thermodynamics also known as entropy to derive a formula for the experimental results derived from the black-body radiation problem.

$$S = k \log W$$

This is Boltzmann's entropy (S), where k is known as the Boltzmann's constant and W is the probability that a particular arrangement of atoms will occur for an element be that a solid, liquid, or gas.

Using Boltzmann's statistical method to calculate entropy, Planck sought a formula to match the results of the black-body experiment. By dividing the total energy (e) in chunks proportional to the frequency (f), he came up with the equation:

$$e = hf$$

where **e** is a chunk of energy, **h** is known as the Planck constant, and **f** is the frequency. Yet, he faced an obstacle; Boltzmann's statistical method demanded the chunks decrease to zero over time. This will nullify his equation and thus defeat its validity. After much struggle, Planck was forced reluctantly to postulate that the energy quantity must be finite. And here comes Planck's incredible insight; if this is correct, it meant that is not possible for an oscillator to absorb or emit energy in a continuous range. It must absorb or emit energy in small indivisible chunks of **e** = **hf** which he called "energy quanta," hence the term quantum mechanics.

Bohr's Quantum Jump

Bohr applied Plank's groundbreaking idea of energy quanta to the atom, the smallest unit of matter. He provided a bold description of the relationship of the atom and light where the electron which rotates around the nucleus will emit or absorb light causing a quantum jump. A quantum jump was therefore a transition between two states; however Bohr was incapable of fully describing it.

This idea was met with skepticism by other scientists who labelled his theory as nonsense, a cheap excuse for not knowing, or too bold, too fantastic to be true. The result was a rift in the physics community with one camp around Bohr believing in the quantum nature of matter and those supporting the classical view. Einstein will soon join the fight in the classical side of the struggle.

Clash of Titans: Quantum Cats and the Uncertainty Principle

By the mid-1920s the new theory about the quantum nature of matter is in shaky ground facing the real prospect of an early demise. It will take two new groundbreaking discoveries to solidify its foundation.

The first came around 1926 when German physicist Werner Heisenberg sought to legitimize Bohr's view by creating a mathematical description of the atom for what is now known as matrix mechanics. This idea was considered too complex to imagine even for the seasoned physicist. Nevertheless, Heisenberg's greatest contribution to the field is his famous uncertainty principle, which we will explore next. A second discovery came from Austrian physicist Erwin Schrödinger who came up with a new description of the atom not as a particle but as a wave. This idea built upon arguments of Louis de Broglie, a French prince who postulated that particles may exhibit wave properties and that duality may be necessary to understand the nature of light (see Figure 1-2).

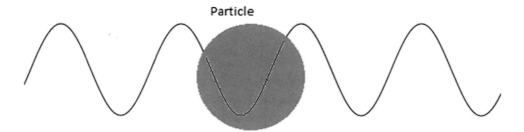

Figure 1-2. *Duality of the nature of the photon. It behaves as both a particle and wave.*

de Broglie used both Einstein's famous equation for energy E = mc^2 and Planck's energy quanta e = hf to find a relation between the wavelength (λ) and the momentum (P) of a photon:

$$E = mc^2 = (m\ c) * c$$

Given that (mc) is the momentum (P) of the photon and c (speed) = f (frequency) * λ (wavelength), the equation becomes:

$$E = (P)(f\lambda)$$

But wait, Planck's relation states that energy E = (h)(f); thus using basic algebra, de Broglie concluded:

$$h * f = P * (f\lambda)$$
$$h = P * \lambda$$
$$\lambda = h / P$$

de Broglie showed that the wavelength of a photon decreases as the momentum increases (see Figure 1-3). By analogy, he proposed that this relation was true not only for photons but for all particles. Given that at the time, the momentum of the electron P = (mass)*(velocity) could be easily determined via experiment; this meant that the wavelength could be calculated from de Broglie's equation! The idea seemed preposterous at the time, as classical physicists knew that the electron was a particle, a discovery made long ago by J. J. Thomson in 1897.

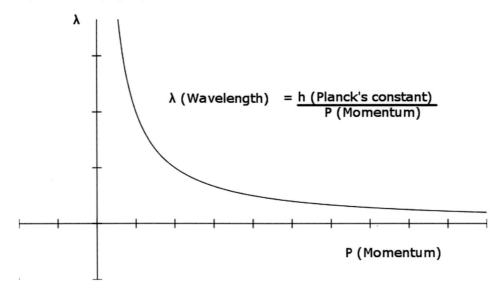

Figure 1-3. *de Broglie relation between the wavelength and the momentum of a photon*

Schrödinger used de Broglie's ideas to find an approach that was more acceptable to the status quo, marking a return to the continuous, visualizable world of classical physics. He was right about his wave function but dead wrong about appeasing the status quo.

Enter the Almighty Wave Function

Schrödinger sought to find a function that could be applied to any physical system for which a mathematical form of energy is known, thus creating his notorious wave function denoted by the Greek symbol ψ (pronounced Psi - see Figure 1-4). The wave function uses Fourier's method of solving equations by expressing any mathematical function as the sum of an infinite series of other periodic functions. This technique is called the method of **eigenvalues** (eigen being the German word for "certain" – a term that is commonplace in quantum physics). Schrödinger wave function was immediately accepted as a mathematical tool of exceptional power for solving problems related to the atomic structure of matter and is considered to be one of the greatest achievements of the twentieth century.

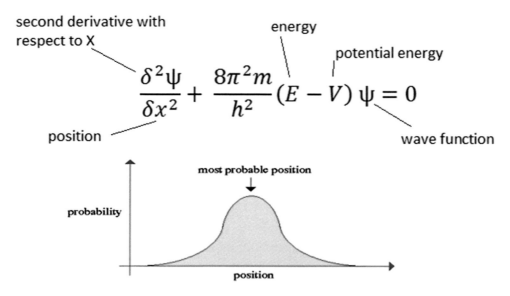

$$\frac{\delta^2 \psi}{\delta x^2} + \frac{8\pi^2 m}{h^2}(E - V)\,\psi = 0$$

Figure 1-4. *Schrödinger famous wave function sought to describe any physical system with known energy*

Bohr and Heisenberg joined forces with Schrödinger given the incredible power of his wave function, but they needed to work out their differences first. It all took place in 1926 at a newly formed institute in Copenhagen where the three giants met to discuss.

Schrödinger rejected the Bohr/Heisenberg concept of discontinuous quantum jumps in the atom structure. He wanted to use his new discovery as a pathway back to the continuous process of physics undisturbed by sudden transitions. He was in fact proposing a classical theory of matter based entirely on waves, even to the point of doubting the existence of particles. Schrödinger proposed that particles are in fact a superposition of waves, a claim that was later proved wrong by Hendrik Lorentz who brought him to his senses, proving that you can't win them all after all. Schrödinger will later waver in his conviction on the importance of wave motion as the source of all physical reality.

Bohr, Heisenberg, and Schrödinger argued relentlessly until the point of exhaustion. Bohr demanded absolute clarity in all arguments, trying to force Schrödinger to admit that his interpretation was incomplete, Schrödinger clinging to his classical view, sometimes bemoaning his work on atomic theory and quantum jumps (something that he probably didn't mean).

Schrödinger loathed Bohr interpretation of the atomic structure. A final piece was required before these two could come to terms on a solid quantum theory.

Probabilistic Interpretation of ψ: The Wave Function Was Meant to Defeat Quantum Mechanics Not Become Its Foundation

Just like when the great rock guitarist Jimi Hendrix heard the tune *Hey Joe*, released a cover, and made it his own, thus creating arguably one of the greatest tune covers, so did the fathers of quantum mechanics. They realized the tremendous power of the wave function and made it their own. A little factoid about this story is that Schrödinger detested Planck's noncontinuous interpretation of energy and heat. He wanted to use his smooth and continuous wave function to defeat Planck's energy quanta. It is hard to believe, but in the 1930s, Planck's discovery was so revolutionary that most physicists thought he was nuts. Nevertheless, just as Hendrix did with that tune, the founders of quantum mechanics will make the wave function theirs.

A breakthrough came from German physicist Max Born, who developed the idea of the wave function as the probability of an electron for a given state to scatter in some direction. Born stated that the probability (P) of the existence of a state is given by the square of the normalized amplitude of the wave function, that is, $P = |\psi|^2$. This was groundbreaking at the time as Born claimed no more exact answers; all we get in atomic theory are probabilities. This brand new idea took Bohr interpretation of the atom in an entirely new direction (see Figure 1-5).

Hydrogen Ground State

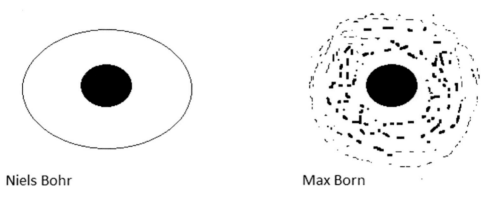

Niels Bohr   Max Born

Figure 1-5. *Bohr vs. Max Born probabilistic view of the wave function*

The Quantum Cat Attempts to Crash Born's Probabilistic Party

As Born's idea about the probabilistic nature of ψ gained traction, Schrödinger through his wave function was being misused, and that originated the famous thought experiment that will be later known as the quantum cat, a story that you probably heard of. In the experiment, Schrödinger sought to rebuff Born's probabilistic interpretation of ψ. It goes like this: a live cat is placed in a box with a radioactive source that triggers the release of a hammer that breaks a flask with poison that will kill the cat. Assuming a 50% probability of radioactive decay per hour, after one hour the mechanism will be triggered, thus killing the cat. Schrödinger claimed that according to Born's interpretation, quantum theory will predict that after one hour, the box would contain a cat that is neither dead nor alive but a mixture of both states, a superposition of both wave functions. Schrödinger thought this was ridiculous and would create a paradox. Yet today, this so-called paradox is used to teach about quantum probabilities and superposition of states.

This is the genius of superposition; as soon as the box is opened, the superimposed wave functions collapse into a single one making the cat dead or alive – thus the act of observation resolves the impasse. Yet another incredible insight will come from Heisenberg pondering about a certain amount of uncertainty about the position of a particle in the atomic structure championed by Bohr.

Uncertainty Principle

Heisenberg pondered about how the position of a particle cannot be known in Bohr's atom. After much reflection, in a moment of clarity, he realized that to know where a particle is, you have to look at it, and to look at it, you have to shine a photon of light on it. However, when you do this, it disturbs the particle position; thus the act of observing a particle changes its location. Heisenberg called this idea the uncertainty principle.

To study the problem, Heisenberg devised a hypothetical experiment using a microscope firing gamma rays, which carry high momentum and low frequency, toward a passing electron to be observed. With Bohr's help, the goal was to describe a quantitative relationship by estimating the imprecision on a simultaneous measurement of the position and momentum. The imprecision of the position was found to be close to the wavelength of the radiation being used, $\Delta X \sim \lambda$.

Similarly, the imprecision of the momentum of the electron is close to the momentum of the photon used to illuminate the particle, $\Delta P \sim h/\lambda$. Note that from the de Broglie equation it is known that the momentum of the photon (P) = h (Planck constant)/ λ (wavelength). Heisenberg showed that multiplying both inequalities, the product will always be greater or equal to h.

$$\Delta X * \Delta P \geq \lambda * h/\lambda$$
$$\Delta X * \Delta P \geq h$$

This is Heisenberg uncertainty principle (HUP) which formally states: "The uncertainty of a simultaneous measurement of the momentum and position is always greater than a fixed amount and close to Planck's constant h."

There is a simple experiment physicists commonly use to show the uncertainty principle in action. It's called the *single slit experiment*, and it goes as follows: A laser beam is fired through a single vertical wide slit and is reflected in a projection screen. What we see with the wide slit is exactly what we suspect, a dot projected on the screen. Now, if we make the width of the slit narrower and narrower, the sides of the dot start to get narrower too. Nevertheless, at around 1/100 of an inch, the uncertainty principle kicks in, and the direction of the beam becomes uncertain, according to Heisenberg. Thus now we observe the light to spread becoming wider and wider! Sounds crazy, how can the light become wider if we are making the slit narrower! It is extremely nonintuitive, but that's how things work.

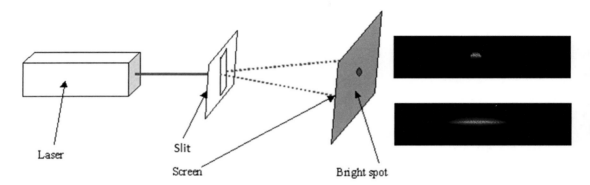

Laser

Slit

Screen

Bright spot

Figure 1-6. *Single slit experiment used to show the uncertainty principle in action*

The uncertainty principle is extremely important because it unifies the rift between Schrödinger and Bohr laying down the foundation of the modern quantum theory. That is, the electron is a particle, as Bohr postulated, but we don't know exactly where it is, as the uncertainty principle states (Heisenberg). Lastly, the probability of finding it is given by the wave function (Schrödinger/Born). Thus there is a duality in the nature of the electron, both as a particle and wave. With all this, a rock-solid view of quantum mechanics emerges that will later be known as the Copenhagen interpretation.

Interference and the Double Slit Experiment

Interference is another incredible property of quantum mechanics, one that makes you think what in the world is going on behind the scenes of our reality. The great physicist Richard Feynman once said about interference: The essentials of quantum mechanics could be grasped from an exploration of interference and the double slit experiment.

It is well known that at the beginning of the nineteenth century, there was a debate raging about the nature of light. Some like Newton claimed it was made of particles; others postulated that it behaved like waves. Thus in 1801, Thomas Young came up with the double slit experiment in an attempt to settle things up: In the experiment, a beam of light is aimed at a barrier with two vertical slits. After the light passes through the slits, the resulting pattern is recorded on a photographic plate. When one slit is covered, a single line of light is displayed, aligned with whichever slit is open. Common sense and intuition tells us that when both slits are open, the resulting pattern would display as two lines of light, aligned with the slits. Incredibly this is not the case. What occurs in practice is that light passing through the slits and displayed on the photographic plate is separated into multiple lines of brightness and darkness in varying degrees (see Figure 1-7).

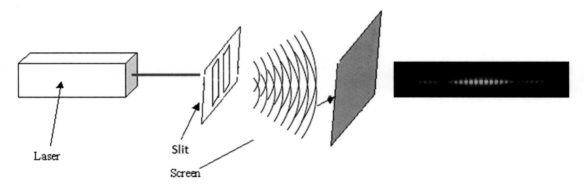

Laser

Slit

Screen

Figure 1-7. Double slit experiment by Thomas Young

This mind-bending result perplexed physicists who hypothesized that interference is taking place between the waves and particles going through the slits. If the beam of photons is slowed enough to ensure that individual photons are hitting the plate, one might expect there to see two lines of light (a single photon going through one slit or the other and ending up in one of two possible light lines). However that is not true. What happens is that somehow the light is doing the impossible: each photon not only goes through both slits but also simultaneously traverses every possible trajectory en route to the target (a principle called interference).

The fact that events like interference, which seem impossible, can occur at the atomic scale baffled the greatest minds at the time. Yet soon, this new theory will face its biggest challenge from the titan of physics, Albert Einstein.

Einstein to Bohr: God Does Not Throw Dice

If you are involved in science, or even if you aren't, you probably heard the famous phrase by Einstein "God does not throw dice." It was coined during a series of letters exchanged with Bohr about the nature of quantum mechanics. Bohr believed the concepts of space-time do not apply at the atomic level. Einstein, on the other hand, was a firm believer in the fabric of space-time and thought this idea could be extended to the atomic scale. This was essentially the root of the disagreement between the two.

Einstein postulated that the properties of an atomic particle could be measured without disturbing it, an idea that goes against the Bohr/Heisenberg interpretation. The two giants faced in a gathering of the greatest physicists of the time in Brussels in 1927 where Einstein sought to prove once and for all that uncertainty does not rule reality.

Einstein challenged Bohr to a series of thought experiments to disprove the uncertainty principle. In round one, Einstein devised a box that he thought will be able to register the precise moment a particle of light was emitted from a small opening in the side of the box and at the same time measure its weight (see Figure 1-7).

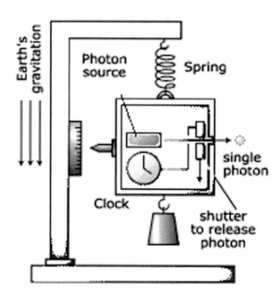

Figure 1-8. *Einstein's experimental box to disprove the uncertainty principle*

In the thought experiment in Figure 1-7, the box has a light source with a clock designed to measure the precise time a photon is emitted. At the same time, the box hangs from a spring with a weight at the bottom and corresponding measuring device. The idea was simple: weight the box before and after the photon is emitted and at the same time register the precise time using the clock. The energy levels could be easily calculated using Einstein's own equation $E = mc^2$. Things didn't look good for the uncertainty principle at that point. If the experiment was correct, the uncertainty principle will be disproven and quantum theory defeated.

Bohr got to work immediately trying to persuade Einstein that if his box works it would mean the end of physics. Bohr prevailed at the end by stating that Einstein forgot to take his own theory into account, as clocks are affected by gravity yielding uncertainty at the time of measurement. He proved the following uncertainty calculation $\Delta E \, \Delta t \geq h$ using Einstein's equation and the red shift formula. Given (Δp) uncertainty of the momentum and (Δq) uncertainty of the position:

$$\Delta p \, \Delta q \geq h \qquad\qquad (1\text{-}1)$$

The uncertainty of the momentum (Δp) is given by $\Delta p \leq t \, g \, \Delta m$; then we have:

$$t \, g \, \Delta m \, \Delta q \geq h \qquad\qquad (1\text{-}2)$$

From the redshift formula and principle of time dilation:

$$\Delta t = c^{-2} \, g \, t \, \Delta q \qquad\qquad (1\text{-}3)$$
$$\Delta E = c^2 \, \Delta m \qquad\qquad (1\text{-}4)$$

Now, multiply (1-3) and (1-4) to obtain (1-5):

$$\Delta E \, \Delta t = g \, t \, \Delta m \, \Delta q \qquad\qquad (1\text{-}5)$$

Finally, comparing (1-5) and (1-2), we obtain an inequality for the uncertainty principle $\Delta E \, \Delta t \geq h$. With this result, round one goes to Bohr; however this will not be the end of it. Einstein believed in a complete picture of physical reality, and the uncertainty principle stood in his way. He will come back with a bigger challenge.

Bohr to Einstein: You Should Not Tell God What to Do

God does not throw dice was Einstein unshakable principle. The firm belief that reality exists independent of one's self. When Einstein wrote to Bohr that god does not throw dice, he replied that he should not tell god what to do. This set the stage for a second struggle between the two while trying to figure out what holds the nucleus together. By the mid-1930s, the time around which this took place, both general relativity and quantum theory are widely accepted as the strongest ideas to explain how the world works. Round two focuses in the most paradoxical aspect of quantum theory – the idea that atomic particles remain connected to one another even at great distances.

Entanglement and the EPR Paradox: Spooky Action at a Distance

In the beginning, light was thought to behave as a wave, but Einstein proved that it also showed particle behavior also known as photons. The same was true about atoms. They behaved both as particles and waves depending on the measuring instrument being used. Furthermore, both conditions were necessary to obtain a complete picture, an idea that Bohr called complementarity.

So how is matter to be understood in the face of these two contradictions? Bohr believed that the atom, as it is, existed outside our perception. This was more than Einstein could accept as he believed on the idea of space-time at the foundation of all physical reality and wanted to extend this concept to the atomic level. Bohr, on the other hand, thought space and time were meaningless and reality was unknowable, and all that we had were phenomena.

It is around this time that Einstein issues a second and final challenge to Bohr. In a paper written with colleagues Podolsky and Rosen, Einstein postulates the question: Does quantum mechanics provide a complete view of physical reality? He proposes a thought experiment where two particles emitted from the same source have common properties and become separated. It should be possible then to measure the first particle and obtain information about the second one without disturbing it. The purpose of the experiment was to demonstrate absurdities in Bohr's view of particles behaving differently based on measuring device. According to quantum mechanics, a measurement on the first particle will influence the other across time and space.

Now, imagine if the particles were to be separated across very large distances (e.g., from one side of the universe to the other). This will create a paradox by violating a fundamental principle of science: the principle of cause and effect. The idea that all events in reality have a cause and effect, and events cannot be transmitted faster than the speed of light, the ultimate speed limit in the universe. Einstein called this principle local causality or locality for short. This paradox will be known as the Einstein-Podolsky-Rosen or EPR paradox.

As soon as Bohr got news of the paper, all work was abandoned immediately. The challenge had to be answered. Bohr was reluctant at first in his reply, but finally did so by claiming that both particles ought to be considered as a single system. In other words both particles become entangled, with space and time being meaningless in such system. Therefore a picture of the atomic world was unknowable.

Einstein called the effects of entangled particles over large distances "Spooky action at a distance." The disagreement between the two was never resolved. Nevertheless, a breakthrough to settle things up came in 1965 by physicist John Bell.

Bell's Inequality: A Test for Entanglement

Bell proposed a set of inequalities to provide experimental proof of the existence of local hidden variables. Formally, Bell's inequality theorem states: *No physical theory of local hidden variables can ever reproduce all of the predictions of quantum mechanics.*[1] Mathematically, it is given by the formula:

$$C_h(a,c) - C_h(b,a) - C_h(b,c) \le 1,$$

$$C_h(a,b) = E\big(A(a,\lambda), B(b,\lambda)\big) = \int_\Delta A(a,\lambda) B(b,\lambda) p(\lambda) d\lambda$$

There is an easy way to understand this very important theorem using simple statistical averages. Consider photon polarization (the oscillation of light in a specific plane) at three different angles A = 0, B = 120, and C = 240 degrees (see Figure 1-8).

---

[1]John Bell, Speakable and Unspeakable in Quantum Mechanics, Cambridge University Press, 1987, p. 65.

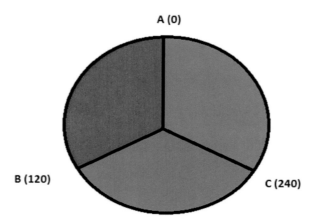

Figure 1-9. *Polarization of light at three angles*

According to Bell's theorem, if reality is independent of observation, then a photon has definite simultaneous values for these three polarization settings, and they must correspond to the eight cases shown in Table 1-2.

Table 1-2. *Permutation Table for Photon Polarization at Three Angles*

Count	A(0)	B(120)	C(240)	[AB]	[BC]	[AC]	Sum	Average
1	A+	B+	C+	1(++)	1(++)	1(++)	3	1
2	A+	B+	C−	1(++)	0	0	1	1/3
3	A+	B−	C+	0	0	1(++)	1	1/3
4	A+	B−	C−	0	1(−−)	0	1	1/3
5	A−	B+	C+	0	1(++)	0	1	1/3
6	A−	B+	C−	0	0	1(−−)	1	1/3
7	A−	B−	C+	1(−−)	0	0	1	1/3
8	A−	B−	C−	1(−−)	1(−−)	1(−−)	3	1

Now ask the simple question: If we measure the polarization at any angle, **what is the probability that the polarization at any neighbor will be the same as the first?** Also calculate the sum and average of the polarizations. In Table 1-2, neighbor polarization is represented by the columns AB, BC, and AC. The + and − signs in columns A, B, and C indicate either positive or negative polarizations at the given angles. Note that there are eight possible permutations, described by the column count. Thus if we find the same polarization (the same sign) for two neighbors, then we record a 1 as well as the sign in columns AB, BC, or AC. This is required to calculate the sum and the average for the respective row in the permutation table.

Now, if a polarization exists independent of measurement (local causality), as Einstein advocates, then the probability of that polarization must be $\geq 1/3$. On the other hand, if Bohr is correct, and reality is defined by the act of observation, then the probability of polarization will be $< 1/3$. This is at the heart of Bell's inequality. Bell does not take sides; it does not say that either is correct but provides the means of finding the truth by experimentation. As a matter of fact, in 1982, French physicist Alain Aspect created an experiment that proved once and for all that Bohr was right all along.

EPR Paradox Defeated: Bohr Has the Last Laugh

In Aspect's experiment, a laser beam irradiates a calcium source producing a pair of photons traveling in opposite directions simultaneously. The photons pass through a polarization filter which only allows a photon polarized in the same plane to pass. It the photon passes, the result is recorded in a measuring device in both sides. Finally, the measuring devices are attached to a counter that registers the results of many interactions (see Figure 1-9).

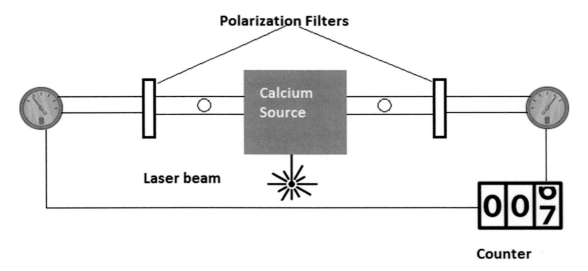

Figure 1-10. Aspect's experiment to test Bell's inequalities - stage 1

If both polarization filters are calibrated in the same direction, Aspect observed a correlation between the pairs of photons. They would either pass or be blocked at the same time. This correlation agreed with Einstein's view of the photon having its polarization property predefined at the moment of emission from the source, not at the moment of measurement as quantum mechanics predicted.

On the other hand, if the polarization settings of the filters are different, then we should expect a certain minimum percentage of photons to either pass or be blocked. Here is where Bell's inequality comes into play (as shown in Table 1-2 of the previous section):

- If the percentage of photons passing through or being blocked is greater than or equal to the expected minimum, then Bell's inequality is preserved and the photon polarization is defined at the moment of emission (the victory goes to Einstein and quantum mechanics is defeated).

- On the other hand, if the percentage is less than the expected minimum, Bell's inequality is violated and quantum physics is correct. The polarization is defined at the moment of measurement (Bohr wins and quantum mechanics is saved).

Aspect performed measurements of many pairs of photons at different polarization settings. The results were astounding: the measurements violated Bell's inequality; thus it was impossible for the polarization to be predefined at the moment of emission. Quantum mechanics was correct! The photons appeared to have chosen a common polarization at the moment of measurement. Could there be some sort of unknown signal between the photons telling them to pick a common value at the moment of measurement?

Einstein's theory of relativity says that no signal can travel faster than the speed of light, the ultimate speed limit in the universe. He called this apparent simultaneous signal spooky action at a distance. Aspect wanted to put this assertion to the test in a second stage of his experiment.

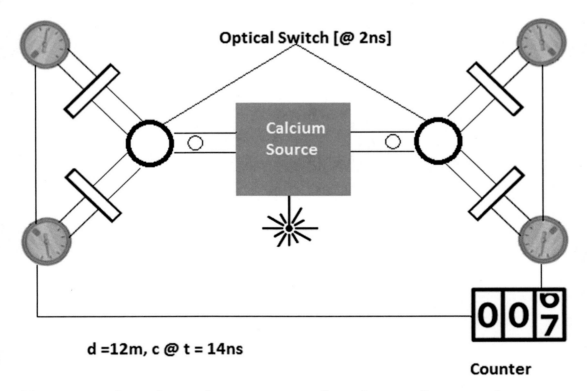

Figure 1-11. *Aspect's experiment to put spooky action at a distance to the test*

In a second stage to his experiment, Aspect uses two optical switches that fork into two separate polarization filters attached to a measuring device each (see Figure 1-10). As before, all measuring devices are attached to a counter to gather results:

- The optical switch is designed to send the photon in one of two directions at an extremely fast rate: 2 nanoseconds or 2 ns.

- The distance between both ends of the experiment was 12 m. It takes the speed of light (traveling at $3*10^8$ meters/second) 40 nanoseconds (ns) to go from one end of the experiment to the other.

Now, if no signal can travel faster than the speed of light, as Einstein's relativity postulates, it should take more than 40 ns from one photon to tell the other what polarization value to choose. Because the optical switch changes at a faster rate (2 ns), the correlation between the photons should not hold. That is, the photons should not be able to choose the same polarization at the moment of measurement (no spooky action at a distance). On the other hand, if the correlation holds, things get extremely weird as some sort of signal is being transmitted to both photons faster than the speed of light.

20

Incredibly, the correlation held in perfect agreement with quantum mechanics, thus proving once and for all that the polarization value was chosen simultaneously by both photons at the moment of measurement faster than the speed of light. The implications were mind blowing as the distance between the photons could have been infinitely grater (e.g., for one end of the universe to the other) or even scarier, across time: from the present into the past or vice versa!

Reality Playing Tricks on Us: Is Everything Interconnected?

Aspect's experiment proves that quantum correlations exist and that if we are to explain them, not just accept them, then we must be bound to admit that some actions occur faster than the speed of light. If that is hard to digest for some, things get even weirder. In a TV interview for the BBC, physicist John Bell said: "There is nothing we can do with this, for example, we cannot send messages or information faster than the speed of light, a fact that is also predicted by quantum mechanics. It seems as nature is playing a trick on us: extraordinary things happen behind the scenes which we cannot use."

At the end, Bohr and Einstein never resolved their differences. They both passed away but their legacy endures. Reading through their fascinating lives, one can't help but wonder: How would have Bohr felt by looking at the results of Alain Aspect experiment proving that he was right all along? Would he have felt happy at his triumph over Einstein? Was all this about the struggle of two egocentric geniuses trying to prove who the better man is? What do you think? I choose to believe that this was a struggle for the advancement of science. All in all, the ultimate winner over the clash of these two titans was humanity.

CHAPTER 2

Quantum Computing: Bending the Fabric of Reality Itself

Semiconductors have come a long way since the days of the vacuum tube. It is hard to believe that the transistor today is around 14 nanometers in size (i.e., close to a molecule). In this chapter you will learn about the origins of quantum computing starting with the fate of the transistor. It seems that the semiconductor process and the transistor are in a collision course with the laws of physics. Next, an in-depth look at the basic component of a quantum computer: the qubit including the strange effects of superposition, entanglement, and qubit manipulation using logic gates. Furthermore, qubit design is an important topic, and this chapter describes the leading prototypes by major IT companies including pros and cons of each.

You will also learn about how quantum computers stack against traditional ones at this point. Things are a little rough for quantum right now, but that is about to change in the next few years. Still, quantum computers face a few pitfalls inherit to the theory of quantum mechanics: they are fragile and error prone; find out why. This chapter also discusses the very interesting quest toward the so-called quantum supremacy. The battle is fierce between IT giants with no winner in sight. Another topic of discussion is the controversial field of quantum annealing and the difference with the standard quantum gate approach used all over this book.

The chapter ends with the path toward universal quantum computation including efforts by all major vendors: in the short term, expect to see quantum computers in the data center. In the long term, the future looks bright with significant resources being poured into fields such as aerospace, medicine, artificial intelligence, and others. The race is getting global. Let's get started.

© Vladimir Silva 2018
V. Silva, *Practical Quantum Computing for Developers*, https://doi.org/10.1007/978-1-4842-4218-6_2

The Transistor Is in a Collision Course with the Laws of Physics

Out of curiosity, have you ever looked inside your home PC to see what it is made of? It is basically a silicon motherboard full of all kinds of electronic gizmos, and in the center rests the big black square that is the CPU. Depending on what kind of PC you have, there may be multiple CPUs, graphics processing units (GPUs), audio, network cards, and all sorts of modularized components. All these components are made of millions of transistors, the fundamental building block of many electronics. A transistor is essentially tiny switch with an on/off position allowing electrons to either pass or not. This property is in turn used to encode a 0 or 1, the basis of the binary language used by all electronics.

Transistors are combined to create logic gates (see Table 2-1). These gates, in turn, produce the fundamental arithmetic functions: addition, subtraction, multiplication, and division. These simple operations provide all the power we need to run powerful scientific simulations, play games, encrypt data, browse the Web, email friends, you name it.

Table 2-1. *Basic Logic Gates*

Type	Symbol	Description	Truth table		
NOT		Negates the input.	A	~A	
			0	1	
			1	0	
AND		Logical product.	A	B	A AND B
			0	0	0
			0	1	0
			1	0	0
			1	1	1

(continued)

24

Table 2-1. (*continued*)

Type	Symbol	Description	Truth table		

OR — Logical addition.

A	B	A OR B
0	0	0
0	1	1
1	0	1
1	1	1

NAND — Negates the logical product.

A	B	A NAND B
0	0	1
0	1	1
1	0	1
1	1	0

NOR — Negates the logical addition.

A	B	A NOR B
0	0	1
0	1	0
1	0	0
1	1	0

(*continued*)

Table 2-1. (*continued*)

Type	Symbol	Description	Truth table		
XOR		Exclusive OR: The output of a two-input XOR is 1 only, when the two input values are different, and 0 if they are equal.	A	B	A XOR B
			0	0	0
			0	1	1
			1	0	1
			1	1	0

Transistors have given our society tremendous technological advances. They are everywhere: computers, communication devices, medicine equipment, aerospace hardware, and others. Whatever machine you can think of is probably made of transistors, yet the transistor is about to face an impassable barrier: the laws of physics, specifically quantum mechanics.

Five-Nanometer Transistor: Big Problem

Since the 1960s traditional computers have grown exponentially in power, at the same time becoming smaller and smaller. Today, computers are made of millions of transistors, but once a transistor starts to get close to the size of an atom, the bizarre world of quantum mechanics kicks in and all bets are off.

Consider Figures 2-1 and 2-2, showing the semiconductor manufacturing process sizes from 1970 to 2020. From a size of around 10 micrometers in the 1970s, sizes become smaller and smaller (at around 1 micrometer) by the late 1980s. Even scarier, there is a huge dip into the nanometer scale (1 nanometer = 10^{-9} m) from the 1990s until the present and beyond (see Figure 2-2). We are talking about transistors approaching the size of molecules. By 2020 the size of a transistor will be around 5 nanometers. At this scale, the bizarre properties of quantum mechanics may start to wreak havoc in the classical computer.

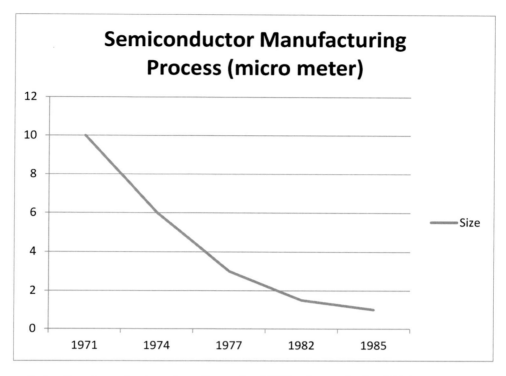

Figure 2-1. *Semiconductor sizes from the 1970s through the 1980s*

Table 2-2. *Semiconductor Size Data for Figure 2-1*

Year	Size in micrometers
1971	10
1974	6
1977	3
1982	1.5
1985	1

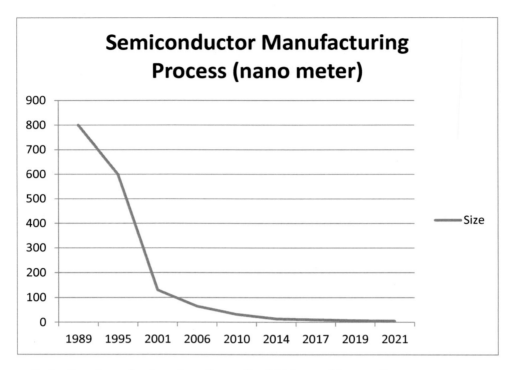

Figure 2-2. *Semiconductor sizes from the 1990s and beyond*

Table 2-3. *Semiconductor Size Data for Figure 2-2*

Year	Size in nanometers
1995	600
2001	130
2010	32
2014	14
2019	7
2021	5

Figure 2-3 shows the size of a transistor by 2020 (around 5 nm) vs. a water molecule (0.275 nm). Unfortunately sizes can't keep getting smaller forever. There is a threshold that will render classical computers useless, and it is called the quantum scale.

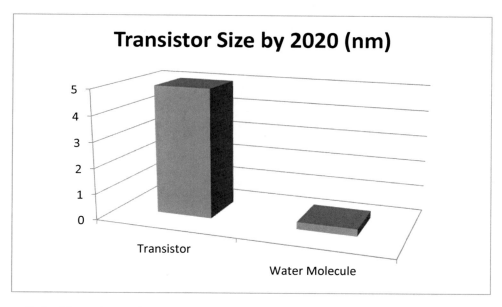

Figure 2-3. *Transistor size against a water molecule*

Quantum Scale and the Demise of the Transistor

Perhaps the demise of the transistor will be an exaggeration. Nevertheless, what is not is the term quantum scale and its effects on it. In physics, quantum scale is the distance where quantum mechanical effects become apparent in an isolated system. This strange boundary lives at scales of 100 nm or less or at a very low temperature. Formally, the quantum scale is the distance at which the action or angular momentum is quantized.

Quantum effects can cascade into the microscale realm causing problems for current microelectronics. The most typical effects are electron tunnelling and interference as shown by the single-double slit experiment.

Electron Tunnelling

Electron tunnelling, also known as quantum tunnelling, is the phenomenon where a particle passes through a barrier that otherwise could not be surmounted at a classic scale. This spells trouble for the transistor and here is why.

Assume that we have a particle with energy E trying to surmount a barrier with potential energy V at the top. According to the classical law of the conservation of energy, the particle needs its energy E > V to pass through, that is, the kinetic energy of the particle must be greater than the potential energy V (see Figure 2-4).

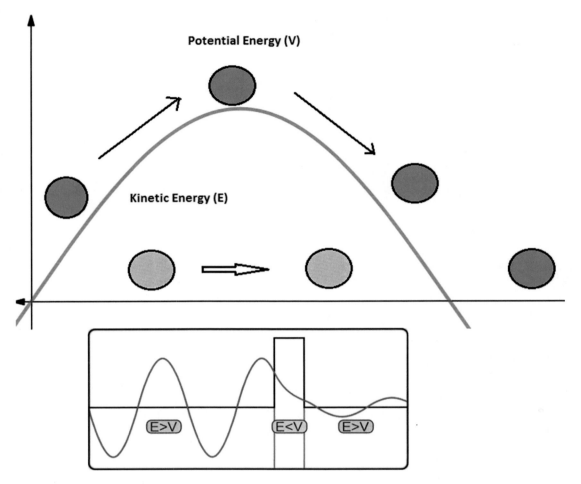

Figure 2-4. *Quantum tunnelling in action*

Note   Electron tunnelling may spell doom for the transistor, yet one's loss is someone else's gain. This important property led to the development of the scanning tunnelling microscope (STM) which had a profound impact on chemical, biological, and materials science research.

Figure 4 shows the effects of classical mechanics as well as quantum tunnelling. According to quantum mechanics, there is a probability that the electron will pass through the barrier even if its kinetic energy is less than the potential energy of the barrier (E < V). This is due to Heisenberg uncertainty principle (HUP). In the previous chapter, you learned about the duality behavior of photons and other particles: both as waves and particles. For waves, Schrödinger's wave function rules, for particles, Niels Bohr described changes in the state of an atom when it gains or losses energy (quantum jumps). The uncertainty principle bridges the gap by introducing the probability of the position and momentum of a particle at a given time.

When a particle such as an electron or photon approaches a barrier (such as a transistor), there is a probability it will go straight through it. This is because its wave function diminishes from sinusoidal to an exponential form,[1] and its solution becomes Equation 2.1.

$$\Psi = Ne^{-\beta x} \tag{2.1}$$

$$P = \exp\left(-\frac{4a\Pi}{h}\sqrt{2m(V-E)}\right) \tag{2.2}$$

where

- ψ is Schrödinger's diminished wave function.

- N is a normalization constant.

- $\beta = \sqrt{2m(V-E)/h^2}$

- m is the mass of the particle.

- V is the potential energy and E is the kinetic energy.

- h is the Planck constant 6.626×10^{-34} m^2kg/s.

- a is the thickness of the barrier.

According to Engel[1], the probability P that the particle will pass through the barrier can be calculated by formula 2.2. Furthermore for quantum tunnelling to occur, the following conditions must be met:

The height of the barrier must be finite and the thickness of the barrier should be thin.

The potential energy of the barrier exceeds its kinetic energy (E<V).

The particle has wave properties suggesting that quantum tunnelling only applies to nanoscale objects such as electrons, photons, etc.

---

[1] Engel, Thomas. Quantum Chemistry and Spectroscopy. Upper Saddle River, NJ. Pearson, 2006. Print.

Let's have some fun by calculating quantum tunnelling probabilities for various barrier sizes of the current semiconductor manufacturing process. The next sections present a series of exercises to visualize this process in detail.

Exercise 1

Calculate the quantum tunnelling probability for the electron using your favorite tool (e.g., an Excel spreadsheet) for formula 2.2 assuming the following values:

- Kinetic energy of the electron E = 4.5 eV.

- Rectangular barrier with potential energy V = 5 eV. Remember that, for quantum tunnelling to occur, E < V.

- Use the size of the barrier provided by the semiconductor manufacturing process in the previous section at the quantum scale; thus size < 100 nm for years 2000 and beyond (use Tables 2-2 and 2-3 from the previous section).

- Don't forget the Planck constant h = 6.626×10^{-34} and the mass of the electron m = 9.1×10^{-31} kg.

Tip   eV is the electron volt, the basic unit of energy in quantum mechanics. 1 eV = 1.6×10^{-19} joules (J). This value is required for unit conversion in the probability calculation.

Solution 1

I have used an Excel spreadsheet to easily calculate values from a table and formula which is included in the source code of this manuscript. Thus pick a cell in your Excel and type formula 2.2. Remember that the part (V-E) must be reduced by multiplication by 1.6×10^{-19} J/eV. Thus formula 2.2 in Excel becomes

EXP(((−4*D5*3.14)/(6.626E−34)) * SQRT(2 * (9.1E−31) * (5 − 4.5) * (1.6E−19)))

In the formula above, cell D5 contains the barrier size, and the rest are the constants π = 3.14, h = $6.626e^{-34}$, m = $9.1e^{-31}$, and 1 eV = 1.6×10^{19} J. With the formula in place, create a new table with the manufacturing year and barrier sizes in nanometers (from Tables 2-2 and 2-3). Finally, make a logic copy of the formula across the cell data for all years and barrier sizes (see Table 2-4).

Table 2-4. *Electron Tunnelling Probabilities for the Semiconductor Process*

Year	Barrier size (m)	Probability
1989	0.0000008	0
2001	1.30E–07	0
2006	0.000000065	6.5829E–205
2010	0.000000032	3.0188E–101
2014	0.000000014	1.053E–44
2017	1.00E–08	3.86767E–32
2019	7.00E–09	1.02616E–22
2021	5.00E–09	1.96664E–16
Beyond	5E–10	0.026876484

What conclusions can be drawn from this data?

- The probability appears to be low, even for the 5 nm manufacturing process coming up in 2021 (1.9e–16). Remember that this value must be multiplied by 100 to obtain a percentage.

- At a barrier size of around 500 picometers (pm), things start to get a little scary. The probability is 0.0268; thus there is 2.68% chance that an electron will pass through the barrier. This means that, for example, if you send some encoded message, 2.68% of the bits will be lost! Not good.

Tip   A picometer (pm) is 1/1000 of a nanometer or $10e^{-12}$ meters.

Exercise 2

Write a tiny program, in your favorite programming language, to calculate the probability as shown in the previous exercise. Verify that the results are the same. Dump the results to standard output as shown in the paragraph below.

Quantum tunnelling probabilities for current semiconductor processes.

2001	1.30e-07	0.000e+00
2010	3.20e-08	2.684e-101
2014	1.40e-08	1.000e-44
2019	7.00e-09	1.000e-22
2021	5.00e-09	1.931e-16
Beyond	5.00e-10	2.683e-02

Solution 2

Listing 2-1 shows a small Java program to calculate the probability for the years and sizes of the current manufacturing process as done in the previous exercise.

Listing 2-1. Java Program to Calculate the Quantum Tunnelling Probability for the Semiconductor Manufacturing Process 2000 and Beyond

```
public class Quantum Tunnelling {

    /** Planck's constant */
    static final double K_PLANK = 6.626e-34;

    /** Mass of the electron (kg) */
    static final double K_ELECTRON_MASS = 9.1e-31;

    /** Electron volt */

    static final double K_EV = 1.6e-19;

    /**
     * Engel's Quantum Tunnelling Probability
     *
     * @param size
     *              Size of the barrier in meters.
     * @param E
     *              Kinetic energy in electron volts (eV).
     * @param V
     *              Potential energy in eV.
     * @return Quantum Tunnelling Probability
     */
```

```java
static double EngelProbability(double size, double E, double V) {
    if (E > V) {
        throw new IllegalArgumentException
            ("Potential energy (V) must be > Kinetic Energy (E)");
    }
    double delta = V - E;
    double p1 = ((-4 * size * Math.PI) / K_PLANK);
    double p2 = Math.sqrt(2 * K_ELECTRON_MASS * delta * K_EV);
    return Math.exp(p1 * p2);
}

/** A simple test for current semiconductor processes */
public static void main(String[] args) {
    try {
        // Barrier sizes for current semiconductor processes (m)
        final double[] SIZES = { 130e-9, 32e-9, 14e-9, 7e-9, 5e-9,
        500e-12 };

        // Dates for display purposes
        final String[] DATES = { "2001", "2010", "2014", "2019",
        "2021", "Beyond" };

        final double E = 4.5; // Kinetic energy of the electron (eV)
        final double V = 5.0; // Potential energy (eV)

        // Display them...
        for (int i = 0; i < DATES.length; i++) {
            double p = EngelProbability(SIZES[i], E, V);

            System.out.println(String.format("%s\t%2.2e\t%2.3e",
            DATES[i], SIZES[i], p));
        }
    } catch (Exception e) {
        e.printStackTrace();
    }
}
}
```

Listing 2-1 defines a function, `EngelProbability`, that takes three arguments: the size of the barrier in meters, the kinetic energy of the particle (E) in eV, and the potential energy (V) in eV. It applies formula 2.1 and returns the probability. The main program then simply loops through an array for the years of the manufacturing process, `String[] DATES = { "2001", "2010", "2014", "2019", "2021", "Beyond"}`, and corresponding sizes: `double[] SIZES = { 130e-9, 32e-9, 14e-9, 7e-9, 5e-9, 500e-12}`. The data is formatted as a table to standard output.

Exercise 3

Plot the data obtained in exercise 1 or 2 into a graph to better visualize the situation. Finally, bravely postulate the year of the demise of the transistor for the current semiconductor manufacturing process!

Solution 3

Spreadsheets are such great tools to manipulate statistical values. The previous data can be plotted in a snap into a cool line graph as shown in Figure 2-5.

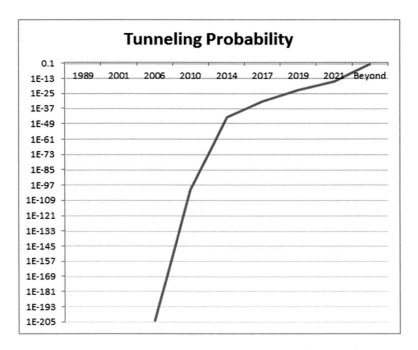

Figure 2-5. *Quantum tunnelling probabilities for the semiconductor manufacturing process*

Now for the grand finale, the demise of the transistor should come around year... I feel skeptical about estimating such year. Something I have learned about quantum mechanics is that everything is ruled by uncertainty. If we assume that a 1% probability for quantum tunnelling is unacceptable for the current manufacturing process, then the data above shows that around 2025, at a barrier size between 1 nanometer and 500 picometers, transistors and therefore all computers may become unusable, although my guess is that transistors will evolve into something else, perhaps something organic or weirder. Nevertheless, it is time to start learning to program a quantum computer just in case.

Now let's look at the next quantum effect causing trouble for the transistor: the uncertainty of the position or momentum shown by basic slit experiments.

Slit Experiments

These experiments were performed many decades ago and are designed as a basic demonstration of the bizarre world of quantum mechanics. They come in many flavors: single slit, double slit, and others. In the single slit experiment, a laser passes through a vertical slit a few inches in width, and it is projected into a surface. The width of the slit can be decreased as desired. As expected we see a dot projected in the surface. Now, if the width of the slit is decreased, the projected dot becomes narrower and narrower; again this is the expected result. But wait, when the width of the slit decreases at about 1/100 of an inch, things get crazy. The dot doesn't become narrower but explodes into a wide horizontal line-like shape. Extremely counterintuitive.

Tip   A more detailed and graphical description of this experiment can be seen in Chapter 1.

Slit experiments are important when talking about transistors because they show the strange effects of quantum mechanics at tiny scales. All in all, Newtonian and relativistic laws of time and space don't make sense at this scale and will create trouble for the transistor.

Possible Futures for the Transistor

Perhaps I am speaking too soon about the demise of the transistor. As a matter of fact, science is already looking ahead at possible alternatives (besides quantum computing). There are some intriguing projects out there to deal with this issue:

- *Molecular electronics*: A field that generates much excitement. It promises to extend the limit of small-scale silicon-based integrated circuits by using molecular building blocks for the fabrication of electronic components. This is an interdisciplinary field that spans physics, chemistry, and materials science.

- *Organic electronics*: A term that sounds fascinating and out of a science fiction movie at the same time. This is a field of materials science concerning the design and application of organic molecules or polymers that show desirable electronic properties such as conductivity. Imagine transistors made of organic materials such as carbon. Not exactly living machines but getting close.

Enter Richard Feynman and the Quantum Computer

The idea of a computational system based on quantum properties comes from Nobel Prize winner physicist Richard Feynman. In 1982 he proposes a "quantum computer" capable of using the effects of quantum mechanics to its advantage.[2] For most of the time since then, interest in quantum computing was mostly theoretical, but things were about to change. In 1995, Peter Shor in his notorious paper "Polynomial-Time Algorithms for Prime Factorization and Discrete Logarithms on a Quantum Computer"[3] proposes a large prime factorization algorithm to run on a quantum computer. This starts a race to create a practical quantum computer when it is proven mathematically that the time complexity (big "O" or execution time) of his algorithm is significantly faster than the current champ of classical computing: the Number Field Sieve.

---

[2]Quantum computation. David Deutsch, Physics World, 1/6/92. A comprehensive and inspiring guide to quantum computing.

[3]Peter W. Shor. Polynomial-Time Algorithms for Prime Factorization and Discrete Logarithms on a Quantum Computer. https://arxiv.org/abs/quant-ph/9508027

The significance of Shor's algorithm is profound in its own right. Consider Figure 2-6 showing the time complexities of the Number Field Sieve against Shor's algorithm.

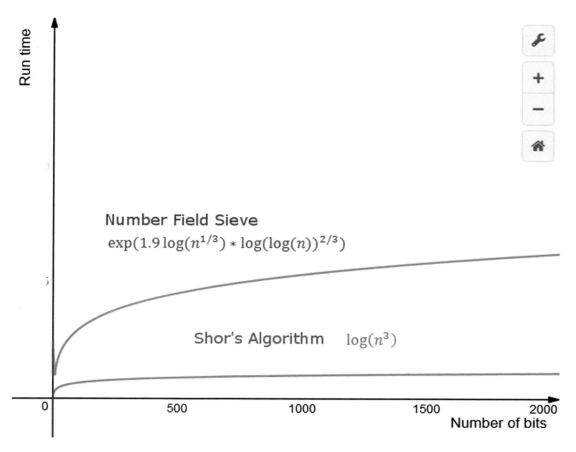

Figure 2-6. *Number Field Sieve vs. Shor's algorithm time complexity*

It has been estimated mathematically that Shor's algorithm could be able to factor a 232-digit integer (RSA-232), one of the current largest integers, in a matter of seconds. Thus a practical quantum computer that can execute Shor's algorithm will render current asymmetric cryptography useless. Keep in mind that asymmetric cryptography is used all over society: at the bank, for example, to encrypt data and accounts, at the Web to browse, communicate, you name it.

But don't rush to the bank, get all your money and put it under the mattress just yet. A practical implementation of this algorithm is decades away right now. This fascinating algorithm is discussed in more detail in a later chapter of this book.

Now back to Feynman and his quantum computer. In a classical computer, the basic unit is the bit (a 0 or 1). In Feynman's computer, the basic unit is the qubit or quantum bit. A unit that is as bizarre as the theory is built upon.

The Qubit Is Weird and Awesome at the Same Time

Just like its classical cousin, the qubit can take a value of either 0 or 1. Physically, qubits can be represented as any two-level quantum systems such as

- The spin of a particle in a magnetic field where up means 0 and down means 1 or

- The polarization of a single photon where horizontal polarization means 1 and vertical polarization means 0. You can make a quantum computer out of light. How weird is that.

In both cases 0 and 1 are the only possible states. Geometrically, qubits can be visualized using a shape called the Bloch sphere, an instrument named after Swiss physicist Felix Bloch (see Figure 2-7).

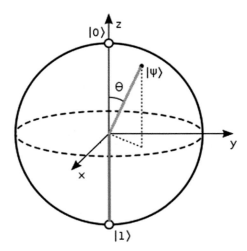

Figure 2-7. *Geometrical representation of a quantum state using the Bloch sphere*

Formally, the Bloch sphere is the geometrical representation in three-dimensional Hilbert space of the pure state of a two-level quantum system or qubit. The north and south poles of the sphere represent the standard basis vectors |0> and |1>, respectively;

these in turn correspond to the spin-up and spin-down of the electron. Besides the basic vectors, the sphere can have something in between; this is called a *superposition* and it is essentially the probability for 0 or 1. The trick is that we can't predict which it will be except at the instant of observation when the probability collapses into a definitive state.

Superposition of States

Imagine if you could flip a coin that could fall not only in a heads or tails position, but in both positions at the same time. Such a coin would be more powerful. Nevertheless there is a catch; the moment you observe this quantum coin, it is forced to take either heads or tails never knowing what position it was in before. This is one reason one needs to be careful when measuring qubits, because they change as soon as observed. All in all, *superposition* is a game changer. Let's see why:

- A 1-bit classical computer can be (or store) in 1 of 2 states at a time: 0 or 1. A 1-qubit quantum computer can be (or store) in 2 states at a time. That is $2^1 = 2$.

- A 2-bit classical computer can store only 1 out of $2^2 = 4$ possible combinations. A 2-qubit quantum computer can store $2^2 = 4$ possible values simultaneously.

Assuming that a byte (8 bits) is the basic unit used to store information in either system, then the number of values that can be stored simultaneously in a quantum computer would be 2^n where n is the number of qubits. Compare this against the storage capacity of a classic computer (shown in Table 2-5) and you realize why qubits are powerful indeed.

Table 2-5. *Qubit Simultaneous Storage Capacity*

Bits/qubits	Classic storage (bytes)	Quantum storage (bits)	Quantum storage (bytes)
4	1	16	2
8	1	256	32
32	4	4294967296	536870912
64	8	1.84467E+19	2.30584E+18

Thus the amount of data that can be stored simultaneously in a quantum computer is astounding, so much so that a new term has popped up out there: *quantum supremacy*. This is the point at which a quantum computer will be able to solve all problems a classical computer cannot. More about this subject will be discussed in a further section of this chapter. But, for now, let's look at the next strange property of the qubit: entanglement.

Entanglement: Observing a Qubit Reveals the State of Its Partner

Long ago, Albert Einstein called quantum entanglement *Spooky action at a distance*. Believe it or not, entanglement has been proved experimentally by French physicist Alain Aspect in 1982. He demonstrated how an effect in one of two correlated particles travels faster than the speed of light!

Tip   Ironically and in a sad twist of faith, humans cannot use entanglement to send messages faster than the speed of light as information cannot travel at such speed. This dichotomy, as well as Aspect's experiment, is explained in more detail in Chapter 1.

If a set of qubits are entangled, then each will react to a change in the other instantaneously, no matter how far apart they are (in opposite sides of the galaxy, e.g., which sounds really unbelievable). This is useful in that, if we measure the properties in 1 qubit, then we can deduce the properties of its partner without having to look. Furthermore, entanglement can be measured without looking through a process called *quantum tomography*. Quantum tomography seeks to determine the state(s) of an entangled set prior to measurement by measurements of the systems coming from the source. In other words, it calculates the probability of measuring every possible state of the system.

> **Note**   Multiqubit entanglement represents a step forward in realizing large-scale quantum computing. This is an area of active research. Currently, physicists in China have experimentally demonstrated quantum entanglement with 10 qubits on a superconducting circuit.[4]

Entanglement is one aspect of qubit manipulation; another mind-bending feature is manipulation via quantum gates.

Qubit Manipulation with Quantum Gates

Gates are the basic building blocks in a quantum computer. Just like their classic counterparts, they operate on a set of inputs to produce another set of outputs. Unlike their cousins however, they operate simultaneously in all possible states of the qubit which makes them really cool and weird at the same time. The basic gates of a quantum computer are:

Measurement Gate

We know that the act of measuring or observing a qubit alters its state. This process is also considered a gate. The measurement gate takes a qubit in a superposition of states as input and spits either a 0 or 1. Furthermore, the output is not random. There is a probability of a 0 or 1 as output which depends on the state the qubit is originally in (see Figure 2-8).

---

[4]Chao Song et al. "10-Qubit Entanglement and Parallel Logic Operations with a Superconducting Circuit." Physical Review Letters. DOI: 10.1103/PhysRevLett.119.180511.

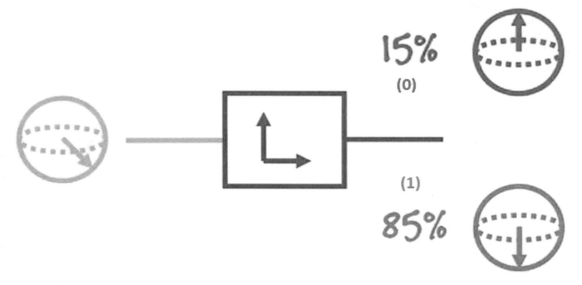

15%

(0)

(1)

85%

Figure 2-8. *Measurement gate and its output probability*

Note that the measurement gate should be the final act on a quantum circuit as quantum mechanics tells us that observing a qubit in the middle of a calculation will collapse its wave function and defeat the parallelism achieved by the superposition of states.

Swap Gate

The swap gate takes 2 qubits and swaps its states as shown in Figure 2-9.

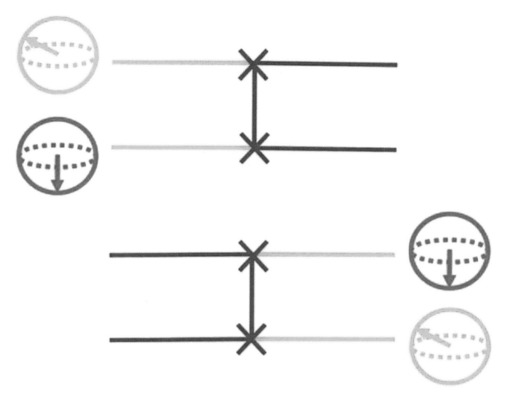

Figure 2-9. *Swap gate in action*

Pauli or X Gate

The Pauli gate is the quantum analog of the classic NOT gate. Formally, it rotates the qubit 180 degrees in the X-axis. Note that the X-axis points outside of the screen as shown in the Bloch sphere in Figure 2-7.

Figure 2-10. *Pauli X gate*

45

Tip   The Pauli gate is named after one of the fathers of quantum physics: Austrian-born Wolfgang Ernst Pauli. In 1945 he won the Nobel Prize in Physics for developing the exclusion principle or Pauli principle which essentially says that no two electrons can exist in the same quantum state.[5] He was highly admired by Albert Einstein and was close friends with giants of quantum mechanics: Niels Bohr and Bernard Heisenberg.

Rotation Gates: Y, Z

Rotation gates over the Y- and Z-axis are known as the Pauli Y and Pauli Z gates, respectively.

- The Pauli Y gate acts on a single qubit. It rotates around the Y-axis of the Bloch sphere by π radians (180 degrees). It maps |0> to i|1> and |1> to –i|0>.

- The Pauli Z gate acts on a single qubit. It rotates around the Z-axis of the Bloch sphere by π radians. It leaves the basis state |0> unchanged and maps |1> to –|1>.

Hadamard Gate (H)

The Hadamard gate acts on a single qubit. It is the combination of two rotations:

1. π over the X-axis

2. π/2 over the Y-axis

The Hadamard gate is the quantum equivalent of the Hadamard matrix, a square matrix whose entries are either +1 or −1 and whose rows are mutually orthogonal.

$$H = \frac{1}{\sqrt{2}} \begin{bmatrix} 1 & 1 \\ 1 & -1 \end{bmatrix}$$

---

[5]Nobel Lecture: Exclusion Principle and Quantum Mechanics Pauli's own account of the development of the Exclusion Principle. www.nobelprize.org/nobel_prizes/physics/laureates/1945/pauli-lecture.html

Tip   The Hadamard transform is useful in data encryption, as well as many signal processing and data compression algorithms.

Controlled (cX cY cZ) Gates

Controlled gates act on 2 or more qubits, where 1 or more qubits act as a control for some operation. For example, the controlled NOT gate (CNOT or cX) acts on 2 qubits and performs the NOT operation on the second qubit only when the first qubit is |1> and otherwise leaves it unchanged.

Toffoli (CCNOT) Gate

This is a controlled gate that operates in 3 qubits. If the first 2 qubits are in the state |1>, it applies a Pauli X (or NOT) on the third qubit; else it does nothing. This gate maps |a,b,c> to |a,b,c + ab>.

The quantum gates shown in Table 2-6 are the basic building blocks of quantum circuits, like the classical logic gates in Table 2-1 are for conventional digital circuits.

Table 2-6. *Basic Quantum Gates*

Gate name	Symbol	Details
Measurement		It takes a qubit in a superposition of states as input and spits either a 0 or 1.
X (NOT)		It rotates the qubit 180 degrees in the X-axis. Maps l0> to l1> and l1> to l0>.
Y	Y	It rotates around the Y-axis of the Bloch sphere by π radians. It is represented by the Pauli matrix: $$Y = \begin{bmatrix} 0 & -i \\ -i & 0 \end{bmatrix}$$ where $i = \sqrt{-1}$ is known as the imaginary unit.

(continued)

Table 2-6. (*continued*)

Gate name	Symbol	Details
Z	$-\boxed{Z}-$	It rotates around the Z-axis of the Bloch sphere by π radians. It is represented by the Pauli matrix: $$Y=\begin{bmatrix} 1 & 0 \\ 0 & -1 \end{bmatrix}$$
Hadamard	$-\boxed{H}-$	It represents a rotation of π on the axis $(X+Z)/\sqrt{2}$. In other words it maps the states: |0> to $(\|0>+\|1>)/\sqrt{2}$ |1> to $(\|0>-\|1>)/\sqrt{2}$
Swap (S)		It swaps 2 qubits with respect to the basis |00>, |01>, |10>, |11>. It is represented by the matrix: $$S=\begin{bmatrix} 1 & 0 & 0 & 0 \\ 0 & 0 & 1 & 0 \\ 0 & 1 & 0 & 0 \\ 0 & 0 & 0 & 1 \end{bmatrix}$$
Controlled (cX cY cZ)		It acts on 2 or more qubits, where 1 or more qubits act as a control for some operation. Its generalized form is described by the matrix: $$C(U)=\begin{bmatrix} 1 & 0 & 0 & 0 \\ 0 & 1 & 0 & 0 \\ 0 & 0 & u_{00} & u_{01} \\ 0 & 0 & u_{10} & u_{11} \end{bmatrix}$$ where U is one of the Pauli matrices σ_x, σ_y, or σ_z.

(*continued*)

Table 2-6. (*continued*)

Gate name	Symbol	Details
Toffoli (CCNOT)		This is a reversible gate, which means that its output can be reconstructed from its input (the states are moved around with no increase in physical entropy). It has 3-bit inputs and outputs; if the first 2 bits are both set to 1, it inverts the third bit; otherwise all bits stay the same. Reversible gates are important because they dissipate less heat. When a logic gate consumes its input, information is lost since less information is present in the output than the input. This loss of information dissipates energy to the surrounding area as heat. In quantum computing Toffoli gates are important because quantum mechanics requires transformations to be reversible and allows more general states (superpositions) of a computation than classical computers.

So a quantum gate manipulates the input of superpositions, rotates probabilities, and produces another superposition as its output. Physically, qubits can be constructed in many ways, with technology companies currently getting into the action in different directions, each design with its own strengths and weaknesses. Let's take a look.

Qubit Design

When it comes to qubit design, only companies with big pockets can get into the race of constructing a practical quantum computer. Due to the weirdness and complexity of quantum mechanics, this is no easy task. In an article for *Science Magazine*[6], writer Gabriel Popkin outlines these efforts from the titans of technology. It seems that all

---

[6]"Scientists are close to building a quantum computer that can beat a conventional one." http://www.sciencemag.org/news/2016/12/scientists-are-close-building-quantum-computer-can-beat-conventional-one

of them want in the action with different designs. Currently, there is no clear winner; nevertheless the race is on. According to Popkin, these are the most common types of qubits:

Superconducting Loops

When an electric current passes through a conductor, some of the energy is lost in the form of heat and light. This is called resistance, and it depends on the type of material; some metals like copper or gold are great conductors of electricity and thus have low resistance. Scientists discovered that the colder the material is, the better conductor of electricity it becomes. Thus the lower the temperature gets, the lower the resistance. Nevertheless, no matter how cold gold or copper gets, it will always show a level of resistance.

Mercury is different however. In 1911, scientists discovered that when cooling down mercury to 4.2 degrees Kelvin (above absolute zero), its resistance becomes zero. This experiment leads to the discovery of the superconductor, a material that has zero electrical resistance at very low temperatures. Since then many other superconducting materials have been found: aluminum, gallium, niobium, and others which show zero resistance at a critical temperature. The great thing about superconductors is that electricity flows without any loss, so a current in a close loop can theoretically flow forever.

Tip   This principle has been proved experimentally when scientists were able to maintain electricity flowing over superconducting rings for years.

In a qubit made of a superconductor loop, a current oscillates back and forth around a loop. A microwave is injected which excites the current into a superposition of states (see Figure 2-11). Let's look at the advantages and disadvantages of such design.

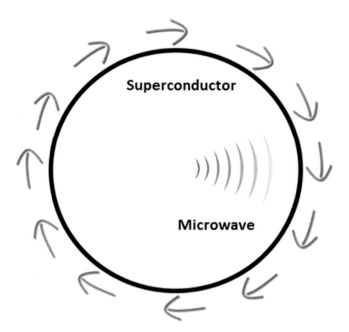

Figure 2-11. *Superconductor loop qubit*

Pros:

- Low error levels (around 99.4% logic success rates)

- Fast, built on existing materials

- Decent number of entangled qubits (9) capable of performing a 2-qubit operation

Cons:

- Low longevity: 0.00005 s. That is, the minimum amount of time a superposition of states can be kept

- Must be kept very cold, at a super frosty –271 °C

This design is used by IBM's cloud platform Q Experience which is the basis for the code used in this book. It is also used by Google and a private venture called Quantum Circuits, Inc. (QCI) which seeks to manufacture a practical quantum computer based on superconductors.

Trapped Ions

An ion trap is a technique for controlling quantum states in a qubit. It uses a combination of electric or magnetic fields to capture charged particles (ions) in a system isolated from the external environment. Lasers are applied to couple qubit states for single operations or coupling between the internal states and the external motional states for entanglement.

Ion traps seek to realize the dream of large-scale universal quantum computing by scaling with *arrays of ion traps*. This technique is also capable of building large entangled states via photonically connected networks of remotely entangled *ion chains* or combinations of these two ideas (see Figure 2-12).

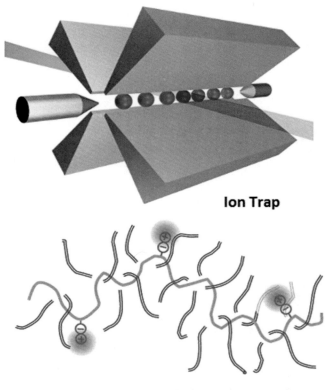

Ion Trap

Entangled 3 qubit Ion chain

Figure 2-12. *Ion traps and chains for large-scale quantum computing*

Let's look at the pros and cons of trapped ions:

Pros:

- High longevity: Experts claim that trapped ions can hold entanglement for up to 1000 s which is huge compared to superconductor loops (0.00005 s).

- Better success rates (99.9%) than superconductors (99.4%). Not much, but still.

- Highest number so far (14) of entangled qubits capable of performing a 2-qubit operation.

Cons:

- Slow operation. Requires lots of lasers.

The top company developing this technology is IonQ located in Maryland, USA.

Silicon Quantum Dots

Intel, the juggernaut of the PC CPU, is spearheading this design. In a silicon quantum dot, electrons are confined vertically to the ground state of a quantum gallium arsenide (GaAs) well, forming a two-dimensional electron gas (2DEG). A 2DEG electron gas is free to move in two dimensions, but tightly confined in the third (see Figure 2-13). This tight confinement leads to quantized energy levels for motion in the third direction which may be of high interest in quantum-based structures.

Tip   2DEG are currently found on transistors made from semiconductors. They also exhibit quantum effects such as the Hall effect, in which a two-dimensional electron conductance becomes quantized at low temperatures and strong magnetic fields.

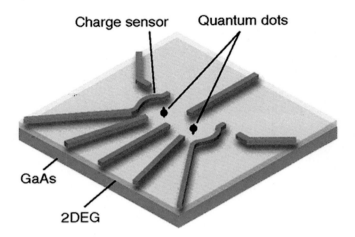

Figure 2-13. *Quantum dot made of gallium arsenide*

The pros and cons of silicon quantum dots are

Pros:

- Stable, built on existing semiconductor materials

- Better longevity than superconductor loops, at 0.03 s

Cons:

- Low number of entangled qubits (2) capable of performing a 2-qubit operation

- Lower success rate than superconductor loops or trapped ions, but still high at 99%

Topological Qubits

Topological qubits seek to eliminate error levels characteristic of quantum computers. Errors are due to the probabilistic nature of quantum mechanics and are described by the longevity or the duration of qubit entanglement. A topological qubit uses two-dimensional quasiparticles called anyons whose paths pass around one another to form braids in a three-dimensional space-time. These braids form the logic gates that make up the computer.

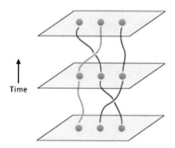

Figure 2-14. *Topological qubit with braids acting as logical gates*

Pros:

- Stable, error-free (longevity doesn't apply)

Cons:

- Purely theoretical at this point, although recent experiments indicate these elements may be created in the real world using semiconductors made of gallium arsenide at a temperature near absolute zero and subject to strong magnetic fields

Microsoft and Bell Labs are some of the companies that support this design.

Diamond Vacancies

Diamond vacancies are locations in the diamond's crystal lattice where there should be a carbon atom but there isn't one. Diamond vacancies seek to harness nanometer-scale atomic defects in diamond materials to function as qubits. It has been observed using atomic force microscopy that the surface of natural diamonds reveals several types of defects. This defect, or vacancy, along with a Nitrogen atom adds an electron to a diamond lattice. The electron quantum spin can then be controlled with a laser as shown in Figure 2-15.

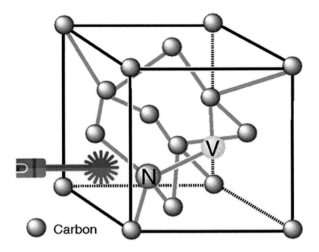

Figure 2-15. *Diamond vacancy qubit*

According to Dirk Englund and colleagues at the MIT School of Electrical Engineering and Computer Science, diamond vacancies solve the perennial problem of reading information out of qubits in a simple way. Diamonds are natural light emitters, and as so, the light particles emitted by diamond vacancies preserve the superposition of states, so they could move information between quantum computing devices. Best of all, they work at room temperatures. No need to cool things down to –272 degrees!

One pitfall of diamond vacancies, says Englund, is that only about 2% of the surface of the diamond has them. Nevertheless researchers are developing processes for blasting the diamond with beams of electrons to produce more vacancies.

Pros:

- High longevity: 10 s.

- High success rate: 99.2%.

- Decent number of entangled qubits (6) capable of performing a 2-qubit operation.

- Qubits operate at room temperature. How incredible is that.

Cons:

- Small number of vacancies in surface materials: about 2%

- Difficult to entangle

All in all, quantum computers have come a long way since the days of Richard Feynman with some of the world's biggest companies looking to cash in. Right now superconductor loops are leading the pack. However there are amazing new designs, such as diamond vacancies, that seek to realize the dream of large-scale quantum computing.

Quantum Computers vs. Traditional Hardware

Quantum computers outmuscle classical hardware for certain tasks. Consider Table 2-7 showing the time complexities for two specific tasks of a quantum vs. a classical computer.

Table 2-7. *Quantum vs. Classical Time Complexities for Certain Tasks*

Task	Quantum	Time complexity	Classical	Time complexity
Search	Grover's algorithm	$\sqrt{n}$	Quick search	n/2
Large integer factorization	Shor's algorithm	$\log(n^3)$	Number Field Sieve	$\exp\left(1.9\log\left(n^{\frac{1}{3}}\right)*\log\left(\log(n)\right)^{\frac{2}{3}}\right)$

For search, Grover's algorithm provides better performance than traditional search. This can have a profound impact in the data center for companies like Google, MS, and Yahoo. Imagine your web search powered by quantum processors in the cloud. We are long ways from that right now, but still. This is one of the reasons big tech companies are investing heavily in developing their quantum platforms.

Another task, and perhaps the main reason why quantum is picking up such steam, is large integer factorization. When Peter Shor came up with his quantum factorization algorithm, he punched a serious crack at the cryptographic security that is at the foundation of our society. Shor's algorithm threatens current encryption systems by factorizing large integers quickly. These integers are used to create the cryptographic keys to encode all data in the Web: bank accounts, business transactions, chats, cat videos, you name it. Shor's algorithm is so fast that, in fact, it could factorize the largest integers of todays in a matter of minutes. Compare that with the current classical champ, the Number Field Sieve, which may take billions of years to factorize such numbers.

Besides search and cryptography, quantum computers can be invaluable tools for simulations, molecular modelling, artificial intelligence, neural networks, and more. Let's see how.

Complex Simulations

Physicists agree that simulations at the atomic level are the field where quantum machines excel. It is the perfect fit after all; a machine built around atoms would be able to simulate quantum mechanical systems with a much greater accuracy than a classical computer. It has been estimated that a quantum computer with a few tens of quantum bits could perform simulations that would take an unfeasible amount of time on a classical computer. For example, the Hubbard model, named after British physicist John Hubbard, which describes the movement of electrons within a crystal, can be simulated by a quantum computer.[7] According to Hubbard, this is a task that is beyond the powers of a classical computer.

Molecular Modelling and New Materials

According to an article in *Science Magazine*, theoretical chemists at the Italian Institute of Technology in Genoa have modelled a molecule of beryllium hydride,[8] a small compound made of two hydrogen and one beryllium atom, in a quantum computer. Not a big deal by today's classical standards but nevertheless a stepping-stone in a future full of hope for new drug discoveries.

Molecular modelling is a virgin new field for quantum machines as physicists and chemists routinely use computers to simulate how atoms and molecules behave. Mathematicians claim that most simulations require massive amounts of computing

---

[7]Hubbard, J. (1963). "Electron Correlations in Narrow Energy Bands". Proceedings of the Royal Society of London. 276 (1365): 238–257. Bibcode:1963RSPSA.276..238H. doi:10.1098/rspa.1963.0204. JSTOR 2414761

[8]"Quantum computer simulates largest molecule yet" By Gabriel Popkin Sep. 13, 2017 available at http://www.sciencemag.org/news/2017/09/quantum-computer-simulates-largest-molecule-yet-sparking-hope-future-drug-discoveries

power. This is especially true for molecular simulations due to the interactions of particles that become exponentially more complex as their numbers increase. Plus the weird laws of quantum mechanics make it hard to calculate the distribution of these electrons within a molecule. Table 2-8 shows some of the current experiments in this field.

Table 2-8. *Quantum Experiments in Molecular Modelling*

Year	Company	Experiment
2016	Google	Researchers at the quantum computing lab in Venice, California, used 3 qubits to calculate the lowest energy electron arrangement of a molecule of hydrogen.
2017	IBM	IBM develops an interactive algorithm to calculate the ground state of specific molecules. Scientists used up to 6 superconductor qubits to analyze hydrogen, lithium hydride, and beryllium hydride by encoding each molecule's electron arrangement into the quantum computer and nudge the molecule into its ground state which they measured and encoded onto a conventional computer.

All in all, molecular modelling has a modest start, but still the future looks bright for chemistry and drug companies. Molecular simulation looks to be a killer application for quantum computing.

Sophisticated Deep Learning

When it comes to deep learning traditional problems fall into three categories: simulation, optimization, and sampling. We have seen in previous sections how a quantum computer excels at simulations, especially at the molecular and atomic levels, but what about optimization? Some optimization problems are not feasible in traditional hardware due to the extreme numbers of interacting variables required to solve them. Examples of these problems include protein folding, space craft flight simulations, and others. Quantum computers can tackle optimization efficiently using a technique called stochastic gradient descent. This is a technique for searching for the best solution among a large set of possible solutions, comparable to finding the lowest point on a landscape of hills and valleys.

Tip   As a matter of fact, a Canadian company called D-Wave already sells commercial quantum computers specifically designed to tackle optimization problems using stochastic gradient descent and other techniques. Some of their customers include defense contractor Lockheed Martin and Google.

Quantum sampling problems fall under the set of computational problems that produce samples from probability distributions. Two classes of sampling problems that demonstrate the power of quantum algorithms are Boson sampling and instantaneous quantum polynomial time sampling. Several small-scale implementations of these two techniques have been performed with quantum optics.

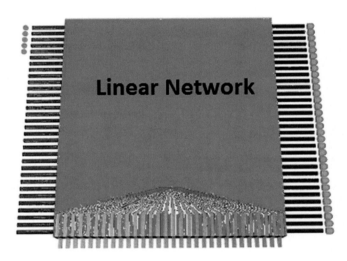

Figure 2-16. *Schematic of the Boson sampling problem*

Figure 2-16 shows a schematic of the Boson sampling problem for a 32 mode instance. Five photons (left) are injected into a linear network that has a scattering matrix (bottom), and all outputs are detected in the Fock basis (right). According to an article in the "Quantum Information" section of *Nature* by A. P. Lund, Michael J. Bremner, and T. C. Ralph, this problem is intractable for classical computers, even for medium-scale systems, such as 50 bosons in 2500 paths. Not even for smaller systems (20 bosons and 400 paths) a feasible classical algorithm is known which can perform this simulation.[9]

---

[9]A. P. Lund, Michael J. Bremner and T. C. Ralph. Quantum sampling problems, BosonSampling and quantum supremacy. Available at www.nature.com/articles/s41534-017-0018-2

Not so much for quantum sampling problems though, which provide a path toward experimental demonstration of the supremacy of quantum algorithms in this field. Deep learning and artificial intelligence are two disciplines that go hand in hand in advanced computation with neural nets being the crown jewel of current research.

Quantum Neural Networks (QNN) and Artificial Intelligence (AI)

Quantum neural networks are the stuff of science fiction more than science fact right now. Nevertheless, the theoretical foundation has been there since the 1990s, and there is wide spread research being done in many directions including

- *The use of quantum information processing to improve existing neural network models*[10]: It is all about boosting existing models with faster and more efficient algorithms. This is a field where quantum computation shines. The motivation for this research is the difficulty to train classical neural networks, especially for big data applications. The hope is that features of quantum computing such as parallelism or the effects of interference and entanglement can be used as resources.

- *Potential quantum effects in the brain*[11]: This path merges quantum physics and neuroscience with a vibrant debate beyond the borders of science. There are pioneers hard at work in the mostly theoretical field of quantum biology which has been gaining momentum by discoveries such as

  - Signs of efficient energy transport in photosynthesis due to quantum effects

  - Reports of "Mag-Lag" effects in MRI scanner patients suggesting that delicate interactions in the brain may be quantum in nature

---

[10]M. Schuld, I. Sinayskiy, F. Petruccione: The quest for a Quantum Neural Network, Quantum Information Processing 13, 11, pp. 2567-2586 (2014).

[11]W. Loewenstein: Physics in mind. A quantum view of the brain, Basic Books (2013).

- *Quantum associative memory*: This is a new algorithm introduced by Dan Ventura and Tony Martinez in 1999.[12] They propose a circuit-based quantum computer that simulates associative memory. The algorithm writes memory states into superpositions, and then it uses a Grover-like quantum search to retrieve the memory state closest to a given input with the ultimate goal being to simulate features of the human brain.

- *Black holes*: Believe it or not, ideas have been proposed about modelling black holes as QNNs and that black holes and brains may store memories in similar ways.[13]

All in all, if a SkyNet-like AI quantum computer is to enslave humanity in the future, chances are that it will be made of some sort of a QNN. It may sound like a joke right now, but giants of science such as Stephen Hawking have warned about this. We should be wise to listen. In the next section we look at the pitfalls that make quantum computers hard to build.

Pitfalls of Quantum Computers: Decoherence and Interference

Decoherence and interference are basic principles in quantum mechanics that cause trouble for large-scale quantum computing.

Decoherence (Longevity)

In quantum mechanics, particles are described by the wave function. A fundamental property of quantum mechanics is called coherence or the definite phase relation between states. This coherence is necessary for the functioning of quantum computers. However, when a quantum system is in contact with its surroundings, coherence decays with time, a process called quantum decoherence. Formally, decoherence is the time

---

[12]D. Ventura, T. Martinez: A quantum associative memory based on Grover's algorithm, Proceedings of the International Conference on Artificial Neural Networks and Genetics Algorithms, pp. 22-27 (1999).

[13]Black Holes as Brains: Neural Networks with Area Law Entropy. Gia Dvali and colleagues. Available at https://arxiv.org/pdf/1801.03918.pdf.

that takes for the superposition of states to disappear and is due to the probabilistic nature of the wave function. It can be viewed as the loss of information from a system into the environment.

Tip   Decoherence was introduced to understand the collapse of the wave function by German physicist H. Dieter Zeh in 1970.[14]

Decoherence can be tested experimentally: quantum mechanics says that particles can be in multiple states (not excited vs. excited or in two different locations) at the same time. Only the act of observation gives a random value for a particular state. If the excitation is measured by the energy levels of the particle (where low energy level means not excited and high energy means excited), when an electromagnetic wave is sent to the particle at a proper frequency, the particle will alternate between high and low energy levels. The state of the particle can then be measured and averaged producing what is called Rabi oscillations. Because the particle is never completely isolated due to atom collisions, electromagnetic fields, or thermal baths, for example, the superposition will stop and the oscillations will disappear.

Thus decoherence gives information about the interaction of a quantum object and its environment and it is crucial for quantum computing. That is, the higher the decoherence (the time it stays in superposition), the higher the quality of the qubit will be. Some qubit designs like superconductor loops have very low longevity and need to be kept at extremely low temperatures (-271 °C) to counter this effect. Others like trapped ions and diamond vacancies have very high longevity and can be kept at room temperatures. Technology companies working in quantum designs face a daunting challenge trying to wrestle with qubit longevity. For a more detailed description of these efforts, see the section under "Qubit Design."

---

[14]Schlosshauer, Maximilian (2005). "Decoherence, the measurement problem, and interpretations of quantum mechanics". Reviews of Modern Physics. 76 (4): 1267–1305. arXiv:quant-ph/0312059 Freely accessible. Bibcode:2004RvMP...76.1267S. doi:10.1103/RevModPhys.76.1267.

Quantum Error Correction (QEC)

Quantum error correction seeks to achieve fault-tolerant quantum computation by protecting information from errors due to decoherence and other environmental noise. When a quantum computer sets up some qubits, it applies quantum gates to entangle them and manipulate probabilities and then finally measures the output collapsing superpositions to a final sequence of 0s or 1s. This means that you get the entire lot of calculations with your circuits done at the same time. Ultimately, you can only measure one output from the entire range of possible solutions. Every possible solution has a probability to be correct so it may have to be rechecked and tried again. This process is called quantum error correction.

In the classical world, error correction is done with redundancy, that is, by creating copies of the data and then assigning probabilities to the possible error conditions and finally comparing the highest probable condition with the original message to determine if an error has occurred. To illustrate this process, consider the following table representing one bit of information:

Message	Redundant copies	Error (1)	Error (1,2)
0	0	1	1
	0	0	1
	0	0	0
Probability		(1/3) = 0.33	(1/3)*(1/3) = 0.11

Let's say that we have a 1 bit message (0) and we create three redundant copies for error correction. Assuming that noisy errors are independent and occur with some probability, it is more likely that the error occurs in a single bit and the transmitted message is three 0s. It is also possible that a double-bit error occurs and the transmitted message is equal to three 1s, but this outcome is less likely. Thus we can use this method to correct the message in case of errors in a classical system. Unfortunately, this is not possible at quantum scales due to the *no-cloning theorem*.

> **Note**   The no-cloning theorem states that it is impossible to create an identical copy of an arbitrary unknown quantum state. It was postulated and proved by Physicist James L. Park in 1970.[15]

The no-cloning theorem creates trouble for quantum computing as redundant copies of qubits cannot be created for error correction. Nevertheless it is possible to spread the information of 1 qubit onto a highly entangled state of several physical qubits. This technique was discovered by Peter Shor with a method of error correcting code by storing the information of 1 qubit onto 9 entangled qubits. However, this scheme only protects against errors of a limited form. Over time several schemes of quantum error correction codes have been developed. The most important are:

The 3-Qubit Code

This is the most basic and the starting point for quantum error correction. This method encodes a single logical qubit into three physical ones with the property that it can correct a single bit-flip error in the Pauli X matrix (σ_x). This code is able to correct errors without measuring the state of the original qubit by using 2 extra qubits to extract what is called *syndrome* information (information regarding possible errors) from the data block without disturbing the original state.

The caveat of this code is that it cannot correct for both bit and phase (sign) flips simultaneously, only a single bit flip. Peter Shor used this method to develop a 9-qubit error correction code.

Shor's Code

This error correction code is based on the 3-qubit code, and it is capable of correcting bit flips, sign flips, or both simultaneously. Shor's code works by encoding 1 single logical qubit into 9 physical qubits using this extra real state to store *syndrome* information about the possible errors. Note that this code can correct errors in a single qubit only. This code tends to be simpler, allowing for more cooperative circuit structures to the physical restrictions of computer architecture. Furthermore, other modern developments in quantum error correction include

---

[15]Wootters, William; Zurek, Wojciech (1982). "A Single Quantum Cannot be Cloned". Nature. 299: 802–803. Bibcode:1982Natur.299..802W. doi:10.1038/299802a0.

- *Bosonic codes*: These try to store error correction information in bosonic modes using the advantage that oscillators have infinitely many energy levels in a single physical system.[16]

- *Topological codes*: These were introduced by physicist Alexei Kitaev with the development of his *toric* code for topological error correction. Its structure is defined on a two-dimensional lattice using error chains which define nontrivial topological paths over the code surface.[17]

All in all, decoherence and quantum error correction are not making things easy for IT companies seeking to realize the dream of large-scale fault-tolerant quantum computing. Nevertheless progress continues at a rapid pace thanks to new qubit designs with high levels of longevity and improved quantum error correction codes. In fact, the pace is so quick that experts in the field have coined a new catchy term for large-scale quantum computing: *quantum supremacy*.

The 50-Qubit Processor and the Quest for Quantum Supremacy

Quantum supremacy is a catchy term indeed. It was coined by physicist John Preskill to describe the point at which a quantum computer could solve problems that classical computers cannot. This is a very potent claim as it requires proof of super-polynomial speedups over their best classic counter parts.

Tip   A super-polynomial speedup is an improvement in the execution of an algorithm above the bounds of a polynomial. For example, an algorithm that runs at $k1n^{c1} + k2n^{c2} + \ldots$ where k and c are arbitrary constants and n is the size of the input is said to be of polynomial time. An algorithm that runs at 2^n where n is the size of the input is said to be of super-polynomial time.

---

[16]Cochrane, P. T.; Milburn, G. J.; Munro, W. J. (1999-04-01). "Macroscopically distinct quantum-superposition states as a bosonic code for amplitude damping". Physical Review A. 59 (4): 2631–2634. doi:10.1103/PhysRevA.59.2631.

[17]A.Y. Kitaev. Quantum Computations: algorithms and error correction. 52:1191, 1997.

Researchers are hard at work on proving quantum supremacy with a few algorithms already in place that provide super-polynomial speedup over the classic champs. The following paragraphs detail a timeline of these efforts.

- *1982*: Richard Feynman, the titan of quantum mechanics, proposes a quantum computer that can take advantage of the atomic principles of superposition, interference, and entanglement. Such a machine will be a game changer.

- *1994*: Mathematician Peter Shor comes up with his notorious factorization algorithm for a quantum computer. The algorithm becomes a sensation when it is estimated that its time complexity crushes the classical super champ (the Number Field Sieve - NFS) by super-polynomial speedups. The algorithm has neither been implemented nor proved experimentally; nevertheless, the genie is out the bottle as excitement grows almost as fast as Shor's algorithm speedup over NFS.

- *2012*: Physicist John Preskill coins the term *quantum supremacy* in the paper "Quantum Computing and the Entanglement Frontier" to formally describe the point in time at which quantum computers will take over. The race is on among the giants of information technology.

- *2016*: Google, the search giant, decides to take on the challenge of proving quantum supremacy by the end of 2017 by constructing a 49-qubit chip that will be able to sample distributions inaccessible to any current classical computers in a reasonable amount of time. The effort fails.

- *2017*: Researchers at IBM T. J. Watson lab perform a simulation of 49- and 56-qubit circuits on a conventional Blue Gene/Q supercomputer at the Lawrence Livermore National Laboratory, increasing the number of qubits needed for quantum supremacy.[18]

---

[18]Edwin Pednault and colleagues. Breaking the 49-Qubit Barrier in the Simulation of Quantum Circuits available online at https://arxiv.org/pdf/1710.05867.pdf.

- *2018*: Skepticism grows on proving quantum supremacy as the pitfalls of quantum computing become more apparent: quantum error correction estimates get as high as 3% of the input on each cycle. Quantum computers are much noisier and error prone compared to their classical counter parts. The Holy Grail becomes a fault-tolerant quantum computer.

Although quantum supremacy is still a long way before a definitive proof, IT insiders predict companies will start seeing returns on their investment into quantum within a few years. Whenever or under which qubit count this so-called quantum supremacy arrives, not even supercomputers will be able to keep up. Believe it or not, there is a company in Canada called D-Wave Systems selling 2000-qubit computers commercially. Although their work remains controversial due to the use of a process called quantum annealing. The next section shows why.

Quantum Annealing (QA) and Energy Minimization Controversy

Quantum annealing sometimes called adiabatic quantum computation (AQC) is a form of quantum computing that relies in the adiabatic theorem to perform calculations. Without getting too technical, here is a list of concepts to better understand this process:

- *Adiabatic theorem*: It was postulated by Max Born and Vladimir Fock in 1928 and states: A quantum mechanical system subjected to gradually changing external conditions adapts its functional form, but when subjected to rapidly varying conditions there is insufficient time for the functional form to adapt, so the spatial probability density remains unchanged.

- *Hamiltonian (H)*: An important concept in quantum mechanics especially so for quantum annealing. In quantum mechanics, a Hamiltonian is an operator corresponding to the total energy of the system in most of the cases. In other words, it is the sum of the kinetic energies of all the particles, plus the potential energy of the particles associated with the system.

Tip   The adiabatic theorem is better understood by a simple example of a pendulum oscillating in a vertical plane. If the support of the pendulum is moved abruptly, the mode of oscillation of the pendulum will change. On the other hand, if the support is moved very slowly, the motion of the pendulum relative to the support will remain unchanged. This is the essence of the adiabatic process: *A gradual change in external conditions allows the system to adapt, such that it retains its initial character.*

In general terms, quantum annealing can be described by the following steps:

1. Find a potentially complicated Hamiltonian whose ground state describes the solution to the problem of interest.

2. Prepare a system with a simple Hamiltonian and initialize to the ground state.

3. Use an adiabatic process to evolve the simple Hamiltonian into to the desired complicated Hamiltonian. By the adiabatic theorem, the system remains in the ground state, so at the end the state of the system describes the solution to the problem.

The pioneer in this form of quantum computing is a company called D-Wave Systems which has sold commercially several quantum computers with fairly large numbers of qubits.

2000 Qubits: Things Are Not As They Seem

Consider the following timeline for a series of quantum systems commercially sold by D-Wave.

- *2007*: D-Wave demonstrates their first 16-qubit hardware.

- *2011*: D-Wave One, a 128-qubit computer sold to Lockheed Martin for 10 million USD.

- *2013*: D-Wave Two, a 512-qubit computer sold to Google for their Quantum Artificial Intelligence Lab seeking to prove quantum supremacy.

- *2015*: D-Wave 2X, breaks the 1000-qubit barrier when sold to an unknown partner.

- *2017*: D-Wave 2000Q, their latest 2000-qubit computer sold to a cybersecurity firm called Temporal Defense Systems for 15 million USD.

It may seem hard to believe that a 2000-qubit quantum computer has already been sold when giants like IBM and Google are just beginning to build 16-qubit systems. Considering that IBM is a company that specializes in large-scale hardware and has the deepest pockets of anybody out there. They would not simply let this happen. Well, the truth is that, in spite of the large number of qubits of the D-Wave 2000Q, it cannot tackle most of the problems that the IBM Q system can.

As a matter of fact, a D-Wave computer can only solve quantum annealing problems, that is, problems solvable by the adiabatic theorem.

Quantum Annealing: A Subset of Quantum Computing

Quantum annealing has been called restrictive by experts in the field generating some controversy due to the following facts:

- Platform like IBM Q use logic gates to control the qubits, while quantum annealing computers don't have logic gates and therefore cannot fully control the state of the qubits.

- D-Wave Systems leverage the fact that their qubits tend to a minimum energy state. They cannot be controlled via quantum gates but their behavior can be predicted by the adiabatic theorem. This makes them good tools for solving energy minimization problems.

- Quantum annealing is used mainly for combinatorial optimization problems where the search space is discrete with a local minimum (e.g., finding the ground state of a disordered magnet/spin glass).[19] QA takes advantage of the notion that all physical systems tend

---

[19]P Ray, BK Chakrabarti, A Chakrabarti "Sherrington-Kirkpatrick model in a transverse field: Absence of replica symmetry breaking due to quantum fluctuations" Phys. Rev. B 39, 11828 (1989).

toward a minimum energy state. To illustrate this, take a hot cup of coffee; when left over the counter for some time, it will start to cool down until it reaches a temperature equal to the surrounding environment. Thus, it tends toward a minimum energy state.

Tip   Mathematical optimization is a technique from the family of local search. It is an iterative method that starts with an arbitrary solution to a problem and then attempts to find a better solution by incrementally changing a single element of the solution. If the change produces a better solution, an incremental change is made to the new solution, repeating until no further improvements can be found.

The question of whether or not the D-Wave QA machine can outmuscle classical computers remains unanswered with several studies going either way: in January 2016, scientists at Google used a D-Wave system to perform a series of tests on *finite-range tunnelling* of a QA solver against *simulated annealing* (SA) and simulated *quantum Monte Carlo* (QMC) on a single-core classical processor.[20] Their results: The QA solver outperformed SA and QMC by a factor of 10^8.

Pretty impressive, however others have said not so fast: Researchers from the Swiss Federal Institute of Technology claimed no quantum speedup for the D-Wave chip but didn't rule out that there could be one in the future.

Figure 2-17 shows the basic inner workings of QA processor such as D-Wave's. It consists of a 2D array of qubits made of superconducting loops which carry an electric current. The qubits act like magnets that can point up, down, or by the properties of quantum mechanics up and down at the same time. Each qubit in the array can interact with others through linkers that can be programmed so that they can lower their energy by either pointing in the same or opposite direction. The idea is to encode a problem by specifying all possible interactions in the chip and solve it by finding the qubit's lowest energy or *ground state*.

---

[20]What is the Computational Value of Finite Range Tunneling? Vasil S. Denchev and colleagues. Google Labs, Jan 2016. Available online at `https://arxiv.org/pdf/1512.02206.pdf`.

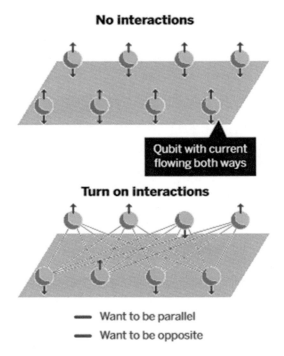

Figure 2-17. *Schematic of a quantum annealing processor*

To find the ground state, the machine starts the array in an entangled state and slowly turns on the interactions. The system then seeks the lowest energy state like a ball rolling thru a valley of peaks to find the deepest point. In classical physics the jiggling of thermal energy drives the ball through the valley to a low point; this is called *thermal annealing*. In quantum mechanics however, the ball can tunnel through low spots to find the lowest even faster. This is the reason why quantum annealing is believed to be faster for problems such as pattern recognition or machine learning.

Thus D-Wave's architecture differs from traditional quantum computers in that it can only solve energy minimization problems. This has created a level of controversy with some at IBM calling it "a dead end." Even the scientists at Google that performed the QA experiment in the D-Wave 2X for instances of Kth order binary optimization quip in their summary that simulated annealing is for the "ignorant or desperate." Adding to the controversy is the fact that D-Wave cannot execute Shor's algorithm because it is not an energy minimization process. Shor's requires what is called a universal quantum computer, a computer that can execute any quantum algorithm.

Universal Quantum Computation and the Future

A universal quantum computer also known as a quantum Turing machine (QTM) is the ultimate quantum machine. It has been defined as an abstract machine capable of seizing all the power of quantum computation. That is, capable of executing any quantum algorithm. Although we are decades away from realizing this dream, a new global race has begun with both major IT industry players and governments pouring significant resources into R&D for these machines.

Google and Quantum Artificial Intelligence

Google has been an early customer of D-Wave and used their machines on a series of optimization experiments whose results showed that quantum annealing can be significantly faster than simulated annealing on a single-core processor. Furthermore, Google has announced that it is developing its own quantum computing technology, a move that makes perfect sense given the amount of resources at their command. Although nothing is available for demonstration at this time, it appears to be a hybrid between IBM's gate-based approach and D-Wave's quantum annealing.

As a matter of fact, Google announced in June 2017 that they are testing a 20-qubit quantum computer with hopes of building a 49-qubit machine by 2018. It seems that they want to give IBM a run for its money in the quest for quantum supremacy. Google has made their quantum wishes clear: artificial intelligence. In the paper "Commercialize Early Quantum Technologies" for Springer Nature, they present the Quantum AI Laboratory with the purpose of building a fault-tolerant quantum machine that can tackle any problem. Google's efforts focus in three key areas of machine learning and artificial intelligence:

- *Simulation*: One of the most anticipated applications is modelling of chemical reactions and materials: stronger polymers for aircraft, improved catalytic converters for cars, more efficient materials for solar, new pharmaceuticals, and breathable fabrics. Quantum computing promises to save untold amounts of money by taking the computer power required to create these materials to the next level. Computational materials are a large industry with a variety of business models built for quantum simulation: pay a subscription for access, consulting, equity exchange in return for quantum-assisted innovations, and others.

- *Optimization*: Optimization problems are difficult to solve with conventional computers. The best classical methods use statistical methods such as energy minimization (thermal annealing). Quantum principles can provide significant speedups by tunnelling thru the thermal barriers in order to find the lowest possible point or best solution. Online recommendations and bidding strategies for advertisements are some of the tasks that require powerful optimization algorithms. In general most machine learning problems can benefit from quantum optimization. Logistics companies, patient diagnosis in health care, and web search companies could achieve tremendous innovation.

- *Sampling*: Mostly related to machine learning tasks such as inference and pattern recognition. Quantum sampling can provide superior performance in probability distribution queries. Not only that, but the massive parallelism achieved by quantum computers can use sampling to provide definitive proof of quantum supremacy.

Google is betting heavily in the future of quantum optimization and risk management, but for now IBM has the advantage with their 20-qubit platform for commercial customers and a 16-qubit free for all cloud platform – *the Q Experience*. One thing is for sure; expect to see cloud-based quantum platforms from every major vendor soon.

Quantum Machines in the Data Center

Qubit design and construction is based on extreme engineering. Because of the bizarre nature of quantum mechanics, qubits are highly susceptible to noise from the environment, error prone due to the principle of decoherence, and in general hard to control and build at a large scale. Thus don't expect quantum computers at the counter of the local Best Buy any time soon. Don't expect either that in the next decades your grandson will be able to buy a quantum computer and plug it in the middle of the family room. Unless a quantum leap in technology is achieved, this is unlikely to happen at all. This is in part because current qubits must be kept at ultrafrosty 0.015 K or around –273 °C to avoid noise from the environment. To have some perspective of this temperature, consider the following table showing average temperatures for different regions of the universe.

Location	Temperature Kelvin	Temperature Celsius
Qubit	0.015	−273
Vacuum of space (the temperature produced by the uniform background radiation or *afterglow* from the Big Bang)	2.7	−270
Average temperature of earth	331	58
Temperatures of the moon at daytime and night	373/100	100/−173

Tip   Kelvin is the primary unit of temperature in physics. Zero Kelvin is defined as the absolute zero or the temperature at which all thermal motion ceases in the classical description of thermodynamics.

What is very likely to happen in the short term is that quantum computers will take over the data center. This means quantum computers will not replace the desktop, but instead perform most of the heavy-duty tasks in the data center such as search, simulations, modelling, and others. Furthermore insiders expect that quantum computers will complement traditional computers offering new types of services such as encryption, scientific intelligence, and artificial intelligence.

So, in a few years, expect the digital assistant in your phone or at home to be powered by a quantum computer. Here is some food for thought: In a decade or so, we will spend most of our time talking to quantum computers.

The Race Is Going Global

Things reach a new level when entire governments get into the action with heavy investments in the field. According to a press release by Digital Single Market, the European Commission is planning to launch a €1 billion flagship initiative starting in 2018 with substantial funding for the next 20 years.[21] This is a follow-up investment in

---

[21]"European Commission will launch €1 billion quantum technologies flagship". Digital Single Market available at https://ec.europa.eu/digital-single-market/en/news/european-commission-will-launch-eu1-billion-quantum-technologies-flagship

addition to the €550 million spent on individual initiatives in order to put Europe at the forefront of what they consider the second quantum revolution.

Furthermore, according to a press release by Alibaba Cloud in July 2015, the Chinese Academy of Sciences (CAS) is teaming up with Alibaba, the largest e-commerce player in China to create the *CAS Quantum Computing Laboratory*. Quantum computing has turned into a global race and the implications will be profound.

Future Applications

There is no limit to the things that can be achieved by the tremendous potential of quantum computing. Here is a list of possible future applications and their impact in our society.

- *Aircraft industry*: Aircraft companies are working in developing and using quantum algorithms for airflow simulations shaving years over their classical counterparts. This will result in more robust and efficient aircraft with low noise and emissions in a fraction of the time.

- *Space applications*: NASA has been toying with the D-Wave system for tasks ranging from optimal structures to optimal packing of payload in a space craft. Other applications include quantum artificial intelligence algorithms and quantum-classical hybrid algorithms.

- *Medicine*: Quantum computing can provide superior molecular simulations resulting in new medicines, lightning fast protein modelling, and faster drug testing. This will reduce the life cycle used to bring medicine to the patient. Next-generation drugs and cancer cures are at our grasp.

These are just some of the future applications of quantum computing. Note that we do not include current breakthroughs such as data encryption and security: quantum factorization and the possibility of defeating asymmetric cryptography are arguably the main reasons quantum computing has picked up so much steam lately. In the next chapter you will get your feet wet with the IBM Q Experience. This is the first quantum computing platform in the cloud that provides real quantum devices for use at our hearts' desire.

Enter the IBM Q Experience: A One-of-a-Kind Platform for Quantum Computing in the Cloud

In this chapter we take a look at quantum computing in the cloud with IBM Q Experience: the first platform of its kind. The chapter starts with an overview of the Composer, the web console used to visually create circuits, submit experiments, explore hardware devices, and more. Next, you will learn how to create your first experiment and submit it to the simulator or real quantum device. IBM Q Experience features a powerful REST API to control the life cycle of the experiment, and this chapter will show you how with detailed descriptions of the end points and request parameters. Finally, the chapter ends with a practical implementation of the official Python library (dubbed IBMQuantumExperience) for Node JS. This custom Node JS library will put your asynchronous Javascript and REST API skills to the test. Let's get started.

IBM has certainly taken an early lead in the race for quantum computing in the cloud. They came up with a really cool platform to run experiments remotely called the Q Experience. But is it just me or the names of these tools make a lot of analogies to music theory? Check this out: the visual editor used to create quantum circuits is called the *Composer*. Not weird enough? The quantum circuits built with the editor are called *scores* (as in a music score), not to mention that visually the editor looks a lot

© Vladimir Silva 2018
V. Silva, *Practical Quantum Computing for Developers*, https://doi.org/10.1007/978-1-4842-4218-6_3

like the written score of a musical composition. I say this because I've been playing the classical guitar for a long time and had an eerie familiarity with a guitar score the first time I looked at the Composer (with the gates looking a lot like music notes). Still think I am crazy? The platform is called the Q Experience; have you ever heard about the Jimi Hendrix Experience? Perhaps the Composer is the score workbook where you will create a great masterpiece for the rest of us to enjoy. Quantum computing does have the power to transform the status quo.

Getting Your Feet Wet with IBM Q Experience

Q Experience is IBM's platform for quantum computing in the cloud, and it is really cool. Let's take a look (All Reprints Courtesy of International Business Machines Corporation, © International Business Machines Corporation):

- Create an account in `https://quantumexperience.ng.bluemix.net/qx/experience`. You will need an email, wait for the approval, and confirm.

- Log in to the web console and navigate to the Composer tab on the top (see Figure 3-1).

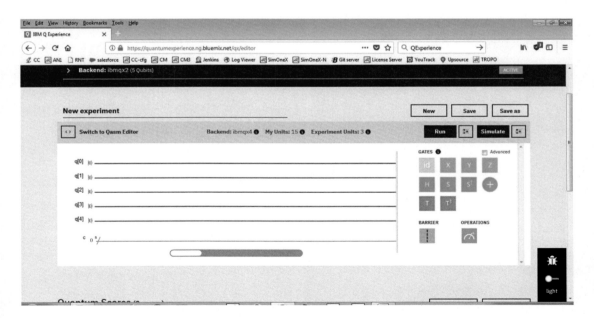

Figure 3-1. *IBM Q Experience main window*

Quantum Composer

The Composer is the visual tool used to create your quantum circuits or scores. At the
top it shows the experiment histogram with the qubits available for use (see Figure 3-2).

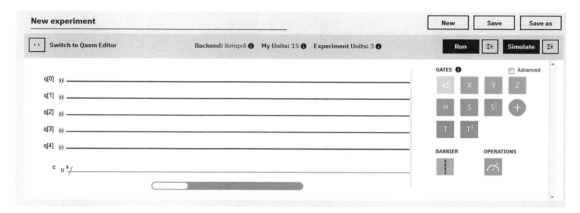

Figure 3-2. *Experiment Composer*

- On the left side of the histogram, we see 5 qubits available from
  processor ibmqx4. They are all initialized to the ground state |0>.
  The line at the bottom is the measurement line where the results of
  the circuit will be collected. Remember that measurement should
  be the last thing done in the circuit as all gate operations execute in
  parallel and in superimposed states.

- On the right side, we have the quantum gates. Drag gates into the
  histogram location of a specific qubit to start building a circuit.

Let's look at the gates and their meaning.

Quantum Gates

The quantum gates supported by IBM Q Experience are described in Table 3-1.

Table 3-1. *Quantum Gates for IBM Q Experience*

Gate	Description
Pauli X X	It rotates the qubit 180 degrees in the X-axis. Maps I0> to I1> and I1> to I0>. Also it is known as the bit flip or NOT gate. It is represented by the matrix: $$X = \begin{bmatrix} 0 & 1 \\ 1 & 0 \end{bmatrix}$$
Pauli Y Y	It rotates around the Y-axis of the Bloch sphere by π radians. It is represented by the Pauli matrix: $$Y = \begin{bmatrix} 0 & -i \\ -i & 0 \end{bmatrix}$$ where $i = \sqrt{-1}$ is known as the imaginary unit.
Pauli Z Z	It rotates around the Z-axis of the Bloch sphere by π radians. It is represented by the Pauli matrix: $$Z = \begin{bmatrix} 1 & 0 \\ 0 & -1 \end{bmatrix}$$
Hadamard H	It represents a rotation of π on the axis $(X+Z)/\sqrt{2}$. In other words, it maps the states: • I0> to $(|0> + |1>)/\sqrt{2}$ • I1> to $(|0> - |1>)/\sqrt{2}$ This gate is required to make superpositions.
Phase $\sqrt{Z}$ S	It has the property that it maps X→Y and Z→Z. This gate extends H to make complex superpositions.
Transposed conjugate of S S†	It maps X→-Y and Z→Z.
Controlled NOT (CNOT) +	This is a 2-qubit gate that flips the target qubit (applies Pauli X) if the control is in state 1. This gate is required to generate entanglement.

(continued)

Table 3-1. (*continued*)

Gate	Description
Phase $\sqrt{S}$ [T]	The $\sqrt{S}$ gate performs halfway of a 2-qubit swap. It is universal such that any quantum multi-qubit gate can be constructed from only sqrt(swap) and single-qubit gates. It is represented by the matrix: $$\sqrt{S} = \begin{bmatrix} 1 & 0 & 0 & 0 \\ 0 & 1/2(1+i) & 1/2(1-i) & 0 \\ 0 & 1/2(1-i) & 1/2(1+i) & 0 \\ 0 & 0 & 0 & 1 \end{bmatrix}$$
Transposed conjugate of T or T-dagger [T†]	Represented by the matrix: $$\sqrt{S} = \begin{bmatrix} 1 & 0 & 0 & 0 \\ 0 & 1/2(1-i) & 1/2(1+i) & 0 \\ 0 & 1/2(1+i) & 1/2(1-i) & 0 \\ 0 & 0 & 0 & 1 \end{bmatrix}$$
Barrier	It prevents transformations across its source line.
Measurement	The measurement gate takes a qubit in a superposition of states as input and spits either a 0 or 1. Furthermore, the output is not random. There is a probability of a 0 or 1 as output which depends on the original state of the qubit.
Conditional [if]	Conditionally apply a quantum operation.
Physical partial rotation (U gates) [U1] [U2] [U3]	U1: It is a one parameter single-qubit phase gate with zero duration. U2: It is a two-parameter single-qubit gate with duration of one unit of gate time. U3: It is a three-parameter single-qubit gate with duration of two units of gate time.
Identity [id]	The identity gate performs an idle operation on the qubit for a time equal to one unit of time.

You can drag gates from the right side of the Composer to create a circuit, or if you prefer to write assembly code, you can switch to the QASM editor mode as shown in Figure 3-3.

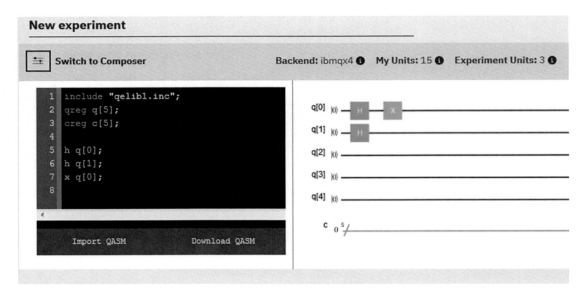

Figure 3-3. *Experiment editor in QASM editor mode*

Tip   QASM is the quantum assembly language, built on top of the OPENQASM platform, and it is used to implement experiments with low-depth quantum circuits. Even though assembler has become something of a lost art, some people may find its raw power more appealing than the Python SDK or even the visual editor.

Now let's take a look at the various quantum processors available for use.

Quantum Backends Available for Use

There are a few quantum processors to choose from for experimentation. Table 3-2 shows the official list ranked by the number of qubits according to the IBM Q Experience backend information site.[1]

---

[1]IBM Q Experience backend information available at https://github.com/QISKit/ibmqx-backend-information

Table 3-2. *Official List of Quantum Backends Available*
for IBM Q Experience Users

Name	Details
Ibmqx2	Code Name: Sparrow Qubits: 5 Online since January 24, 2017
Ibmqx4	Code Name: Raven Qubits: 5 Online since September 25, 2017
Ibmqx3	Code Name: Albatross Qubits: 16 Online since June 2017
Ibmqx5	Code Name: Albatross Qubits: 16 Online since online September 28, 2017 This device is reconfigured version of ibmqx3.

Table 3-2 shows the official list of processors available for use at the time of this
writing, but there is a much interesting way to get an updated list of available machines
in real time using the excellent REST API. This API is described in more detail on the
"Remote Access via the REST API" section in this chapter, but for now let's demonstrate
how to obtain an always up-to-date list of backends using the *Available Backend List*
REST end point:

```
https://quantumexperience.ng.bluemix.net/api/Backends?access_
token=ACCESS-TOKEN
```

Tip   To obtain an access token, see the section "Authentication via API
Token" under "Remote Access via the REST API" of this chapter. Note that an
API token is not the same as an access token. API tokens are used to execute
quantum programs via the Python SDK. Access tokens are used to invoke the
REST API.

The URL in the previous paragraph returns a list of quantum processors in JSON format. This is what it looks like by the time of this writing. Note that your results may be different:

Listing 3-1. HTTP Response from the Backend Information REST API Call

```
[{
  "name": "ibmqx2",
  "version": "1",
  "status": "on",
  "serialNumber": "Real5Qv2",
  "description": "5 transmon bowtie",
  "basisGates": "u1,u2,u3,cx,id",
  "onlineDate": "2017-01-10T12:00:00.000Z",
  "chipName": "Sparrow",
  "id": "28147a578bdc88ec8087af46ede526e1",
  "topologyId": "250e969c6b9e68aa2a045ffbceb3ac33",
  "url": "https://ibm.biz/qiskit-ibmqx2",
  "simulator": false,
  "nQubits": 5,
  "couplingMap": [
    [0, 1],
    [0, 2],
    [1, 2],
    [3, 2],
    [3, 4],
    [4, 2]
  ]
}, {
  "name": "ibmqx5",
  "version": "1",
  "status": "on",
  "serialNumber": "ibmqx5",
  "description": "16 transmon 2x8 ladder",
  "basisGates": "u1,u2,u3,cx,id",
  "onlineDate": "2017-09-21T11:00:00.000Z",
```

```
  "chipName": "Albatross",
  "id": "f451527ae7b9c9998e7addf1067c0df4",
  "topologyId": "ad8b182a0653f51dfbd5d66c33fd08c7",
  "url": "https://ibm.biz/qiskit-ibmqx5",
  "simulator": false,
  "nQubits": 16,
  "couplingMap": [
    [1, 0],
    ...
    [15, 14]
  ]
}, {
  "name": "Device Real5Qv1",
  "status": "off",
  "serialNumber": "Real5Qv1",
  "description": "Device Real5Qv1",
  "id": "cc7f910ff2e6860e0d4918e9ee0ebae0",
  "topologyId": "250e969c6b9e68aa2a045ffbceb3ac33",
  "simulator": false,
  "nQubits": 5,
  "couplingMap": [
    [0, 1],
    [0, 2],
    [1, 2],
    [3, 2],
    [3, 4],
    [4, 2]
  ]
}, {
  "name": "ibmqx_hpc_qasm_simulator",
  "status": "on",
  "serialNumber": "hpc-simulator",
  "basisGates": "u1,u2,u3,cx,id",
  "onlineDate": "2017-12-09T12:00:00.000Z",
  "id": "084e8de73c4d16330550c34cf97de3f2",
```

```
  "topologyId": "7ca1eda6c4bff274c38d1fe66c449dff",
  "simulator": true,
  "nQubits": 32,
  "couplingMap": "all-to-all"
}, {
  "name": "ibmqx4",
  "version": "1",
  "status": "on",
  "serialNumber": "ibmqx4",
  "description": "5 qubits transmon bowtie chip 3",
  "basisGates": "u1,u2,u3,cx,id",
  "onlineDate": "2017-09-18T11:00:00.000Z",
  "chipName": "Raven",
  "id": "c16c5ddebbf8922a7e2a0f5a89cac478",
  "topologyId": "3b8e671a5a3b56899e6e601e6a3816a1",
  "url": "https://ibm.biz/qiskit-ibmqx4",
  "simulator": false,
  "nQubits": 5,
  "couplingMap": [
    [1, 0],
    [2, 0],
    [2, 1],
    [2, 4],
    [3, 2],
    [3, 4]
  ]
}, {
  "name": "ibmqx3",
  "version": "1",
  "status": "off",
  "serialNumber": "ibmqx3",
  "description": "16 transmon 2x8 ladder",
  "basisGates": "u1,u2,u3,cx,id",
  "onlineDate": "2017-06-06T11:00:00.000Z",
  "chipName": "Albatross",
```

```
    "id": "2bcc3cdb587d1bef305ac14447b9b0a6",
    "topologyId": "db99eef232f426b45d2d147359580bc6",
    "url": "https://ibm.biz/qiskit-ibmqx3",
    "simulator": false,
    "nQubits": 16,
    "couplingMap": [
    ...
    ]
}, {
    "name": "QS1_1",
    "version": "1",
    "status": "standby",
    "serialNumber": "QS1_1",
    "description": "20 qubit device v1",
    "basisGates": "SU2+CNOT",
    "onlineDate": "2017-10-20T11:00:00.000Z",
    "chipName": "Qubert",
    "id": "cb141f7bb641b8a10487a6fab8483b86",
    "topologyId": "25197b9b73c4b52ca713ca4d126417b5",
    "simulator": false,
    "nQubits": 20,
    "couplingMap": [
    ...
    ]
}, {
    "name": "ibmqx_qasm_simulator",
    "status": "on",
    "description": "online qasm simulator",
    "basisGates": "u1,u2,u3,cx,id",
    "id": "18da019106bf6b5a55e0ef932763a670",
    "topologyId": "250e969c6b9e68aa2a045ffbceb3ac33",
    "simulator": true,
    "nQubits": 24,
    "couplingMap": "all-to-all"
}]
```

Listing 3-1 shows the current list of available processors which mostly matches the official list from the IBM Q Experience web site. However, there is a lot of extra interesting information about the structural layout of these machines:

- Extra processors and simulators:

  - It looks like there are two remote simulators available for use (ibmqx_qasm_simulator, ibmqx_hpc_qasm_simulator) even though the official documentation mentions only one: ibmqx_qasm_simulator. This information can come in handy when testing complex circuits: more simulators are always a good thing.

  - Rumors of a 20-qubit processor have been swirling around for some time. There is even talk of an upcoming 50-qubit monster processor by the end of 2018. This list seems to confirm the 20-qubit machine at least. But don't get excited just yet; this machine is only available for corporate customers.

- Besides the usual information such as machine name, version, status, number of qubits, and others, there are some terms we should be familiarized with:

  - *basisGates*: These are the physical qubit gates of the processor. They are the foundation under which more complex logical gates can be constructed. Most of the processors in the list use u1, u2, u3, cx, id.

    - Gates u1, u2, u3 are called *partial NOT gates* and perform rotations on axes X, Y, Z by theta, phi, or lambda radians of a qubit.

    - Cx is called the *controlled NOT* gate (CNOT or CX). It acts on 2 qubits and performs the NOT operation on the second qubit only when the first qubit is |1> and otherwise leaves it unchanged.

    - Id is the identity gate which performs an idle operation on a qubit for one unit of time.

- *couplingMap*: The coupling map defines interactions between individual qubits while retaining quantum coherence (or a pure state – imagine a peloton of soldiers breaking step when crossing an old bridge so that the amplitude of their feet hitting the ground does not add up and destroy the bridge). Qubit coupling is used to simplify quantum circuitry and allow the system to be broken up into smaller units.

Now back to the Composer for our first quantum composition.

Opus 1: Variations on Bell and GHZ States

This composition is a weird one. Here we look at two mind-bending quantum experiments used to demonstrate the weirdness of quantum mechanics:

- *Bell states*: They demonstrate that physics are not described by local reality. This is what Einstein called *spooky action at a distance*.

- *GHZ states*: Even stranger than Bell states, GHZ states (named after their creators: Greenberger-Horne-Zeilinger) are the 3-qubit generalization of the Bell states.

Let's look at them in more detail.

Bell States and Spooky Action at a Distance

Bell states are the experimental test of the famous Bell inequalities. In 1964 Irish physicist John Bell proposed a way to put quantum entanglement (spooky action at a distance) to the test. He came up with a set of inequalities which have become incredibly important in the physics community. This set of inequalities is known as Bell's theorem, and it goes something like this.

Consider photon polarization (when light oscillates in a specific plane) at three different angles A = 0, B = 120, and C = 240. Realism says that a photon has definite simultaneous values for these three polarization settings, and they must correspond to the eight cases shown in Table 3-3.

Table 3-3. *Permutations for Photon Polarizations at Three Angles*

Count	A(0)	B(120)	C(240)	[AB]	[BC]	[AC]	Sum	Average
1	A+	B+	C+	1(++)	1(++)	1(++)	3	1
2	A+	B+	C−	1(++)	0	0	1	1/3
3	A+	B−	C+	0	0	1(++)	1	1/3
4	A+	B−	C−	0	1(−−)	0	1	1/3
5	A−	B+	C+	0	1(++)	0	1	1/3
6	A−	B+	C−	0	0	1(−−)	1	1/3
7	A−	B−	C+	1(−−)	0	0	1	1/3
8	A−	B−	C−	1(−−)	1(−−)	1(−−)	3	1

Now Bell's theorem asks: *what is the probability that the polarization at any neighbor will be the same as the first?* We also calculate the sum and average of the polarizations. Assuming realism is true, then by looking at Table 3-3, the answer to the question is the probability must be >= 1/3. This is what Bell's inequality gives: a means to put this assertion to the test. Here is the incredible part: believe it or not, quantum mechanics violates Bell's inequality giving probabilities less than 1/3. This was proven experimentally for the first time in 1982 French physicist Alain Aspect.

Tip   A more detailed description of Aspect's experiment and Bell's inequality is described in Chapter 1, EPR Paradox Defeated: Bohr Has the Last Laugh.

So now let's translate the photon polarization from the preceding text into an experiment that can be run in a quantum computer. In 1969 John Clauser, Michael Horne, Abner Shimony, and Richard Holt came up with a proof for Bell's theorem: the CHSH inequality which formally states

$$S = \langle A,B \rangle - \langle A,B' \rangle + \langle A',B \rangle + \langle A',B' \rangle$$

$$S \leq 2$$

To illustrate this, we have two detectors: Alice and Bob. Given A and A′ are detector settings on side Alice, B and B′ on side Bob, with the four combinations being tested in separate experiments. Realism says that for a pair of entangled particles, the parity table showing all possible permutations looks as shown in the following table:

A	B	1
A	B′	0
A′	B	0
A′	B′	1

In classical realism, the CHSH inequality becomes $|S| = 2$. However, the mathematical formalism of quantum mechanics predicts a maximum value for S of $|S| = 2\sqrt{2}$, thus violating this inequality. This can be put to the test using four separate quantum circuits (one per measurement) with 2 qubits each. To simplify things, let measurements on Alice detector be $A = Z$ and $A′ = X$, and Bob's detector $B = W$ and $B′ = V$ (see Table 3-4). To begin the experiment, a basis Bell state must be constructed which matches the identity (see Figure 3-4):

$$1/\sqrt{2}\left(|00\rangle + |11\rangle\right)$$

The preceding expression essentially means the qubit held by Alice can be 0 or 1. If Alice measured her qubit in the standard basis, the outcome would be perfectly random, either possibility having probability 1/2. But if Bob then measured his qubit, the outcome would be the same as the one Alice got. So, if Bob measured, he would also get a random outcome on first sight, but if Alice and Bob communicated, they would find out that, although the outcomes seemed random, they are correlated.

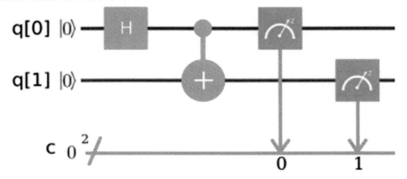

Figure 3-4. *Basis Bell state*

In Figure 3-4, 2 qubits are prepared in the ground state |0>. The H gate creates a superposition of the first qubit to the state $1/\sqrt{2}\,(|00+|10\rangle)$. Next the CNOT gate flips the second qubit if the first is excited, making the state $1/\sqrt{2}\,(|00+|11\rangle)$. This is the initial entangled state required for the four measurements in Table 3-4 (All reprints courtesy of International Business Machines Corporation, © International Business Machines Corporation).

- To rotate the measurement basis to the ZW axis, use the sequence of gates S-H-T-H.

- To rotate the measurement basis to the ZV axis, use the sequence of gates S-H-T'-H.

- The XW and XV measurement is performed the same way as in the preceding text and the X via a Hadamard gate before a standard measurement.

Tip   Before performing the experiment in the Composer, make sure its topology (the number of qubits and target device) in the score is set to 2 over a simulator. Some topologies (like the 5 qubits in a real quantum device) do not support entanglement for qubits 0 and 1 giving errors at design. Note that the target device can be a real quantum processor or a simulator. All in all, as long as you use a simulator, you should be fine.

Table 3-4. *Quantum Circuits for Bell States*

Bell state measurement	Result for 100 shots	

AB (ZW)

c[2]	Probability
11	0.39
10	0.06
00	0.46
01	0.09

AB′ (ZV)

c[2]	Probability
11	0.49
10	0.07
00	0.36
01	0.08

A′B (XW)

c[2]	Probability
11	0.42
10	0.05
00	0.49
01	0.04

A′B′ (XV)

c[2]	Probability
11	0.05
10	0.52
00	0.03
01	0.40

Now we need to construct a table with the results of each measurement plus the correlation probability between A and B <AB>. The sum of the probabilities for the parity of the entangled particles is given by

$$\langle AB \rangle = P(1,1) + P(0,0) - P(1,0) - P(0,1)$$

Remember that the ultimate goal is to determine if S ≤ 2 or |S| = 2; thus by compiling the results of all measurements, we obtain Table 3-5.

Table 3-5. *Compiled Results from the Bell Experiment*

	P(00)	P(11)	P(01)	P(10)	<AB>
AB (ZW)	0.46	0.39	0.09	0.06	0.68
AB' (ZV)	0.36	0.49	0.08	0.07	0.73
A'B (XW)	0.49	0.42	0.04	0.05	0.47
A'B'(XV)	0.03	0.05	0.4	0.52	−0.32

Add the absolute values of column <AB> and we obtain |S| = 2.2. These results violate Bell's inequality (as predicted by quantum mechanics) and are pretty close to the official tests performed on May 2, 2017, by IBM scientists over 8192 shots.[2] How about yours?

Even Spookier: GHZ States Tests

These are named after physicists Greenberger-Horne-Zeilinger who came up with a generalization test for N entangled qubits with the simplest being a 3-qubit GHZ state:

$$|GHZ\rangle = 1/\sqrt{2}\left(|000\rangle - |111\rangle\right)$$

---

[2]IBM Q Experience Bell Tests Results available online at https://quantumexperience.
ng.bluemix.net/proxy/tutorial/full-user-guide/003-Multiple_Qubits_Gates_and_
Entangled_States/002-Entanglement_and_Bell_Tests.html

Note   The importance of the GHZ states is that they show that the entanglement of more than two particles is in conflict with local realism not only for statistical (probabilistic) but also nonstatistical (deterministic) predictions.

In simple terms GHZ states show a stronger violation of Bell's inequality. Let's see how with a simple puzzle: imagine three independent boxes each containing two variables X and Y. Each variable has two possible outcomes: 1 and –1. The question is to find a set of values for X and Y that solves the following set of identities:

```
(1)    XYY = 1
(2)    YXY = 1
(3)    YYX = 1
(4)    XXX = -1
```

For the impatient out there, there is no solution for this. For example, replace $Y = 1$ in (1), (2), and (3), and then multiply them, that is, $(5) = (1) * (2) * (3)$. The set then becomes

```
(1)    X11 = 1
(2)    1X1 = 1
(3)    11X = 1
(4)    XXX = -1
(5)    Multiply (1) (2) (3) and we get XXX = 1
```

There is no solution because identity (4) $XXX = -1$ contradicts identity (5) $XXX = 1$. The scary part is that a GHZ state can indeed provide a solution to this problem, which seems impossible in the deterministic view of classical reality, but nothing is impossible in the world of quantum mechanics, just improbable.

Incredibly, GHZ tests can rule out the local reality description with certainty after a single run of the experiment, but first we must construct a GHZ basis state.

Table 3-6. *Basis GHZ State*

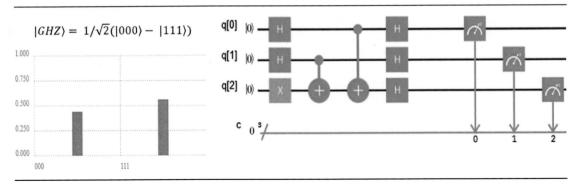

To kick-start the experiment, the basis GHZ state (as well as probability results which should be around half) is shown in Table 3-6:

1. In the basis circuit, Hadamard gates in qubits 1 and 2 put them in superposition |00,01,01,11>. At the same time, the X gate negates qubit 3; thus we end up with the states $1/\sqrt{2}\left(|001\rangle + |101\rangle + |011\rangle + |111\rangle\right)$.

2. The two CNOT gates entangle all qubits into the state $1/\sqrt{2}\left(|001\rangle + |010\rangle + |100\rangle + |111\rangle\right)$.

3. Finally the three Hadamard gates map step 2 to the state ½(|000⟩ − |111⟩).

Now, create the quantum circuits for identities XYY, YXY, XYY, and XXX from the previous section as shown in Table 3-7 (All reprints courtesy of International Business Machines Corporation, © International Business Machines Corporation).

Table 3-7. *Quantum Circuits for GHZ States*

Measurement	Results for 100 shots	

YYX

c[3]	Probability
011	0.34
101	0.23
110	0.23
000	0.20

YXY

c[3]	Probability
011	0.23
101	0.28
110	0.25
000	0.24

XYY

c[3]	Probability
011	0.23
101	0.26
110	0.35
000	0.16

XXX

c[3]	Probability
010	0.25
100	0.32
111	0.22
001	0.21

- For the measurement of X, apply the H gate to the corresponding qubit.

- For each instance of Y, apply the S† (S-dagger), and H gates to the corresponding qubit.

Finally compare the results of the preceding experiment against the official data from IBM Q Experience.[3] How do your results stack up? All in all, the principles of quantum mechanics shown in this section have been challenged by a theory called super determinism which gives a way out.

Super Determinism: A Way Out of the Spookiness. Was Einstein Right All Along?

In an interview for BBC in 1969, physicist John Bell talked about his work on quantum mechanics. He said that we must accept the predictions that actions are transferred faster than the speed of light between entangled particles but at the same time we cannot do anything with it. Information cannot travel faster than the speed of light, a fact that is also predicted by quantum mechanics. As if nature is playing a trick on us. He also mentioned that there is a way out of this riddle through a principle called super determinism.

Particle entanglement implies that measurements performed in one particle affect the other instantaneously, even across large distances (think opposite sides of the galaxy or the universe), even across time. Einstein was an ardent opponent of this theory famously writing to Niels Bohr *God does not throw dice*. He could not accept the probabilistic nature of quantum mechanics, so in 1935, along with colleagues Podolsky and Rosen, they came up with the infamous EPR paradox to challenge its foundation. In the EPR paradox, if two entangled particles are separated by a tremendous distance, a measurement in one could not affect the other instantaneously as the event will have to travel faster than the speed of light (the ultimate speed limit in the universe). This will violate general relativity, thus creating a paradox: Nothing travels faster than the speed of light, that is, the absolute rule of relativity.

---

[3]GHZ States Experiment available online at `https://quantumexperience.ng.bluemix.net/proxy/tutorial/full-user-guide/003-Multiple_Qubits_Gates_and_Entangled_States/003-GHZ_States.html`

Nevertheless, in 1982 the predictions of quantum mechanics were confirmed by French physicist Alain Aspect. He devised an experiment that showed Bell's inequality is violated by entangled photons. He also proved that a measurement in one of the entangled photons travels faster than the speed of light to signal its state to the other. Since then, Aspect's results have been proven correct time and again (details on his experiment is shown in Chapter 1). The irony is that there is a chance that Einstein was right all along and entanglement is just an illusion. It is the principle of super determinism.

Tip   In simple terms super determinism says that freedom of choice has been removed since the beginning of the universe. All particle correlations and entanglements were established at the moment of the Big Bang. Thus there is no need for a faster-than-light signal to tell particle B what the outcome of particle A is.

If true, this loophole will prove that Einstein was right when postulating the EPR paradox and all our hard work in quantum programing is just an illusion. But this principle sounds more like religious dogma (all outcomes determined by fate) than science as Bell argued that super determinism was implausible. His reasoning being that freedom of choice is effectively free for the purpose at hand due to alterations introduced by a large number of very small effects. Super determinism has been called untestable as experimenters would never be able to eliminate correlations that were created at the beginning of the universe. Nevertheless this hasn't stopped scientists on trying to prove Einstein right and particle entanglement an illusion. As a matter of fact, there is an experiment hard at work to settle things up and is really inventive. Let's see how.

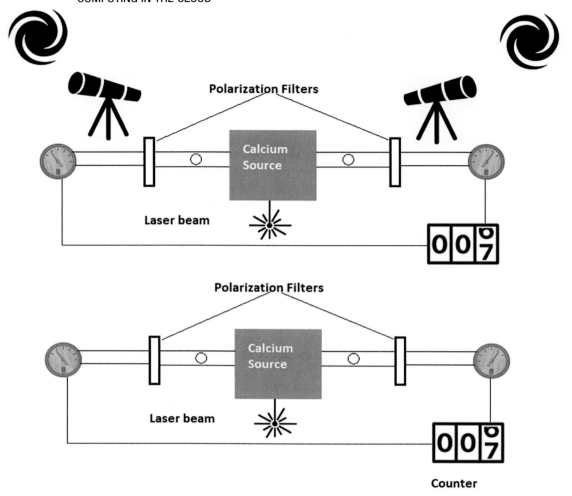

Figure 3-5. *Bell's inequality experiment using cosmic photons vs. the*
standard test

Figure 3-5 shows the standard Bell's inequality test experiment (at the bottom) and a
variation of the experiment using cosmic photons (at the top) by Andrew Friedman and
colleagues at MIT.[4]

---

[4]Jason Gallicchio, Andrew S. Friedman, and David I. Kaiser. Testing Bell's Inequality with Cosmic
Photons: Closing the Setting-Independence Loophole. Available online at http://web.mit.edu/
asf/www/Papers/Gallicchio_Friedman_Kaiser_2014.pdf

Tip   For a full description of the standard Bell's inequality test, see Chapter 1, EPR
Paradox Defeated: Bohr Has the Last Laugh.

Friedman and colleagues came up with a novel variation of the standard Bell
experiment using cosmic rays. The idea is to use real-time astronomical observations
of distant stars in our own galaxy, distant quasars, or patches of the cosmic microwave
background, to essentially let the universe decide how to set up the experiment instead
of using a standard quantum random number generator. That is, photons from the
distant galaxies are used to control the orientation of the polarization filters just prior to
the arrival of entangled photons.

If successful, the implications would be groundbreaking. If the results from such
experiment do not violate Bell's inequality, it would mean that super determinism could
be true after all. Particle entanglement will be an illusion, and signal transfer between
entangled particles could not travel faster than light as predicted by relativity. Einstein
will be right and there is no spooky action at a distance.

Luckily for us, fans of quantum mechanics, no such thing has happened so far. Keep
in mind that Friedman and colleagues are not the only team getting it on the action.
There are multiple teams trying to crack this riddle. As a matter of fact, most of their
results agree with quantum mechanics. That is, their results violate Bell's inequality. So
it seems that the rift created by Einstein and Bohr in their struggle between relativity
and quantum mechanics long ago is alive and well. My money is in quantum mechanics
though. Moving on, the next section shows how IBM Q Experience can be accessed
remotely via its slick REST API.

Remote Access via the REST API

Q Experience features a relatively unknown REST API that handles all remote
communications behind the scenes. It is used by the current Python SDKs:

- *QISKit*: The Quantum Information Software Kit is the de facto access
  tool for quantum programming in Python.

- *IBMQExperience*: A lesser known library bundled with QISKit that
  wraps the REST API in a Python client.

In this section we peek inside IBMQExperience and look at the different REST end points for remote access. But first, authentication is required.

Authentication

To invoke any REST API call, we must first obtain an access token. This will be the access key to invoke any of the calls in this section. Note that the access token is not the same as the API token (the API token is used to execute quantum programs in Python). There are two ways of obtaining an access token:

- *Using your API token*: To obtain the API token, log in to the IBM Q Experience console and follow the instructions in the following section.

- *Using your account username and password*: Let's see how this is done using REST.

Tip   To obtain your API token, log in to the IBM Q Experience console, click your username ➤ *My Account*, and then click the *Advanced* tab on the upper right. Finally click *Generate* and then *Copy API Token* (see Figure 3-6). Always keep your token secure.

Figure 3-6.   *Obtain your API token from the console*

Authentication via API Token

- **HTTP Method:** POST

- **URL:** `https://quantumexperience.ng.bluemix.net/api/users/`
  `loginWithToken`

- **Payload:** {"apiToken": "YOUR_API_TOKEN"}

Authentication via User-Password

- **HTTP Method:** POST

- **URL:** `https://quantumexperience.ng.bluemix.net/api/users/`
  `login`

- **Payload:** {"email": "USER-NAME", "password": "YOUR-PASSWORD"}

The response for both methods is

```
{
  "id": "ACCESS_TOKEN",
  "ttl": 1209600,
  "created": "2018-04-15T20:21:03.204Z",
  "userId": "USER-ID"
}
```

Where *id* is your access token, *ttl* is the time to live (or expiration time) in milliseconds, and *userId* is your user id. Save the access token and the user id for use in this section. Note that when your session expires, a new access token needs to be generated.

List Available Backends

This call returns a JSON list of all available backends and simulators in IBM Q Experience:

- **HTTP Method:** GET

- **URL:** `https://quantumexperience.ng.bluemix.net/api/`
  `Backends?access_token=ACESS-TOKEN`

Request Parameters

Name	Value
access_token	Your account access token

HTTP Headers

Name	Value
x-qx-client-application	Defaults to qiskit-api-py

Response Sample

The response content type for all API calls is application/json. The next paragraph shows
the partial result of a call to this end point. Note that this end point will return both real
processors and simulators.

```
[{
        "name": "ibmqx2",
        "version": "1",
        "status": "on",
        "serialNumber": "Real5Qv2",
        "description": "5 transmon bowtie",
        "basisGates": "u1,u2,u3,cx,id",
        "onlineDate": "2017-01-10T12:00:00.000Z",
        "chipName": "Sparrow",
        "id": "28147a578bdc88ec8087af46ede526e1",
        "topologyId": "250e969c6b9e68aa2a045ffbceb3ac33",
        "url": "https://ibm.biz/qiskit-ibmqx2",
        "simulator": false,
        "nQubits": 5,
        "couplingMap": [
                [0, 1],
                [0, 2],
                [1, 2],
```

```
            [3, 2],
            [3, 4],
            [4, 2]
        ]
    },..]
```

The most important keys from the preceding response are described in Table 3-8.

Table 3-8. *Available Backend Response Keys*

Key	Description
Name	The name id of the processor to be used when executing code against it.
Version	A string or positive integer probably used to track changes to the processor.
Description	This is probably a description of the hardware used to build the chip. You may see things like: • 5 transmon bowtie • 16 transmon 2x8 ladder Note: a transmon is defined as a type of noise-resistant superconducting charge qubit. It was developed by Robert J. Schoelkopf, Michel Devoret, Steven M. Girvin, and their colleagues at Yale University in 2007.[5]
basisGates	These are the physical qubit gates of the processor. They are the foundation under which more complex logical gates can be constructed.
nQubits	The number of qubits used by the processor.
couplingMap	The coupling map defines interactions between individual qubits while retaining quantum coherence. It is used to simplify quantum circuitry and allow the system to be broken up into smaller units.

---

[5]J. Koch et al., "Charge-insensitive qubit design derived from the Cooper pair box," Phys. Rev.
A 76, 04319 (2007), doi:10.1103/PhysRevA.76.042319, arXiv:0703002

Get Calibration Information for a Given Processor

This call returns a JSON list of the calibration parameters for a given processor in Q Experience. These parameters are documented in detail in the IBMQX backend information site.[6]

- **HTTP Method:** GET

- **URL:** `https://quantumexperience.ng.bluemix.net/api/Backends/NAME/calibration?access_token=ACCESS-TOKEN`

Request Parameters

Name	Value
access_token	Your account access token.

HTTP Headers

Name	Value
x-qx-client-application	Defaults to qiskit-api-py (a default value for the official client although I suspect it can be anything).

Response Sample

Qubits are highly sensitive to error and environmental noise. Calibration information gives an overview of the quality of the qubits inside the processor. Listing 3-2 shows a simplified response of the calibration parameters for ibmqx4. Some of the most remarkable parameters are

- **gateError:** This is the error rate of a qubit gate operation at a given time.

- **readoutError:** This is the error rate of a qubit readout operation at a given time.

---

[6]ibmqx-backend-information available online at `https://github.com/QISKit/ibmqx-backend-information/tree/master/backends`.

Tip   Qubit quality evaluation involves four stages (operations): preparation, memory, gates, and readout. Error rates are calculated at the gate and readout stages to track the quality of the qubit. This is the information returned by this API call. Note that after usage qubits must be reset (cooled down) to a basis state.

Listing 3-2. Simplified Response for the Calibration Parameters If ibmqx4

```
{
      "lastUpdateDate": "2018-04-15T10:47:03.000Z",
      "qubits": [{
            "gateError": {
                  "date": "2018-04-15T10:47:03Z",
                  "value": 0.0012019552727863259
            },
            "name": "Q0",
            "readoutError": {
                  "date": "2018-04-15T10:47:03Z",
                  "value": 0.049
            }
      }, ...
],
      "multiQubitGates": [{
            "qubits": [1, 0],
            "type": "CX",
            "gateError": {
                  "date": "2018-04-15T10:47:03Z",
                  "value": 0.03024023736391171
            },
            "name": "CX1_0"
      },...
]}
```

The information in Listing 3-2 can be seen under the IBM Q Experience console
Devices tab on the main menu (see Figure 3-7). Get the calibration information via REST,
and compare it against the web console (Reprint courtesy of International Business
Machines Corporation, © International Business Machines Corporation).

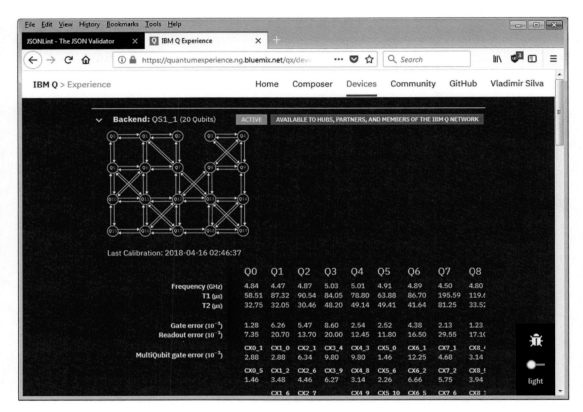

Figure 3-7. *Calibration information reported by the web console*

Get Backend Parameters

This call returns a JSON list of the backend parameters for a given processor in Q
Experience. Some of these parameters include

- Qubit cooldown temperature in Kelvin degrees: For example, I got
  0.021 K for ibmqx4 – that is, a super frosty –459.6 °F or –273.1 °C.

- Buffer times in ns.

- Gate times in ns.

- Other quantum specs documented in more detail at the backend information site.[7]

The request type and end point URL are

- **HTTP Method:** GET

- **URL:** `https://quantumexperience.ng.bluemix.net/api/Backends/NAME/parameters?access_token=ACCESS-TOKEN`

Request Parameters

Name	Value
access_token	Your account access token

HTTP Headers

Name	Value
x-qx-client-application	Defaults to qiskit-api-py

Response Sample

Listing 3-3 shows a simplified response for ibmqx4 parameters in JSON.

Listing 3-3. Simplified Response for ibmqx4 Parameters

```
{
    "lastUpdateDate": "2018-04-15T10:47:03.000Z",
    "fridgeParameters": {
        "cooldownDate": "2017-09-07",
        "Temperature": {
            "date": "2018-04-15T10:47:03Z",
```

---

[7]IBM Q Experience backend information available online at `https://github.com/QISKit/ibmqx-backend-information`.

```
                    "value": 0.021,
                    "unit": "K"
            }
    },
    "qubits": [{
            "name": "Q0",
            "buffer": {
                    "date": "2018-04-15T10:47:03Z",
                    "value": 10,
                    "unit": "ns"
            },
            "gateTime": {
                    "date": "2018-04-15T10:47:03Z",
                    "value": 50,
                    "unit": "ns"
            },
            "T2": {
                    "date": "2018-04-15T10:47:03Z",
                    "value": 16.5,
                    "unit": "µs"
            },
            "T1": {
                    "date": "2018-04-15T10:47:03Z",
                    "value": 45.2,
                    "unit": "µs"
            },
            "frequency": {
                    "date": "2018-04-15T10:47:03Z",
                    "value": 5.24208,
                    "unit": "GHz"
            }
    },..]
```

Get the Status of a Processor's Queue

This call returns the status of a specific quantum processor event queue.

- **HTTP Method:** GET

- **URL:** `https://quantumexperience.ng.bluemix.net/api/Backends/`
  `NAME/queue/status`

Request Parameters

It seems strange but this API call appears not to ask for an access token.

HTTP Headers

Name	Value
x-qx-client-application	Defaults to qiskit-api-py

Response Sample

For example, to get the event queue for ibmqx4, paste the following URL in your browser:

`https://quantumexperience.ng.bluemix.net/api/Backends/ibmqx4/queue/status`

The response looks like `{"state":true,"status":"active","lengthQueue":0}` where

- **state:** It is the status of the processor. If alive, true else false.

- **status:** It is the status of the execution queue – active or busy.

- **lengthQueue:** It is the size of the execution queue or the number of simulations waiting to be executed.

Tip   When you submit an experiment to IBM Q Experience, it will enter an execution queue. This API call is useful to monitor how busy the processor is at a given time.

List Jobs in the Execution Queue

This call returns a list of jobs in the processor execution queue.

- **HTTP Method:** GET

- **URL:** `https://quantumexperience.ng.bluemix.net/api/` `Jobs?access_token=ACCESS-TOKEN&filter=FILTER`

Request Parameters

Name	Value
access_token	Your account access token.
filter	A result size hint in JSON. For example, {"limit":2} returns a maximum of two entries.

HTTP Headers

Name	Value
x-qx-client-application	Defaults to qiskit-api-py

Response Sample

Listing 3-4 shows the response format for this call. The information appears to be a historical record of experiment executions containing information such as status, dates, results, code, calibration, and more.

Listing 3-4. Simplified Response for the Get Jobs API Call

```
[{
    "qasms": [{
        "qasm": "...",
        "status": "DONE",
        "executionId": "331f15a5eed1a4f72aa2fb4d96c75380",
        "result": {
            "date": "2018-04-05T14:25:37.948Z",
            "data": {
```

```
                          "creg_labels": "c[5]",
                          "additionalData": {
                                   "seed": 348582688
                          },
                          "time": 0.0166247,
                          "counts": {
                                   "11100": 754,
                                   "01100": 270
                          }
                 }
             }
         }
     }],
     "shots": 1024,
     "backend": {
          "name": "ibmqx_qasm_simulator"
     },
     "status": "COMPLETED",
     "maxCredits": 3,
     "usedCredits": 0,
     "creationDate": "2018-04-05T14:25:37.597Z",
     "deleted": false,
     "id": "d405c5829274d0ee49b190205796df87",
     "userId": "ef072577bd26831c59ddb212467821db",
     "calibration": {}
}, ...]
```

Note   Depending on the size of the execution queue, you may get an empty
result ([]) if there are no jobs in queue or a formal result as shown in Listing 3-4.
Whatever the case make sure the HTTP response code is 200 (OK).

Get Account Credit Information

When an account is created, each user is assigned a number of default execution credits (15) which are spent when running experiments. This call lists your credit information.

- **HTTP Method:** GET

- **URL:** `https://quantumexperience.ng.bluemix.net/api/users/USER-ID?access_token=ACCESS-TOKEN`

Tip   The user id can be obtained from the authentication response via API token or user-password. See the "Authentication" section for details.

Request Parameters

Name	Value
access_token	Your account access token.

HTTP Headers

Name	Value
x-qx-client-application	Defaults to qiskit-api-py

Response Sample

Listing 3-5 shows a sample response for this call.

Listing 3-5. Credit Information Sample Response

```
{
     "institution": "Private Research",
     "status": "Registered",
     "blocked": "None",
     "dpl": {
          "blocked": false,
```

```
        "checked": false,
        "wordsFound": {},
        "results": {}
    },
    "credit": {
        "promotional": 0,
        "remaining": 150,
        "promotionalCodesUsed": [],
        "lastRefill": "2018-04-12T14:05:09.136Z",
        "maxUserType": 150
    },
    "additionalData": {
    },
    "creationDate": "2018-04-01T15:36:16.344Z",
    "username": "",
    "email": "",
    "emailVerified": true,
    "id": "",
    "userTypeId": "...",
    "firstName": "...",
    "lastName": "..."
}
```

List User's Experiments

This call lists all experiments for a given user id.

- **HTTP Method:** GET

- **URL:** https://quantumexperience.ng.bluemix.net/api/users/
  USER-ID/codes/lastest?access_token=ACCESS-TOKEN&includeExe
  cutions=true

Request Parameters

Name	Value
USER-ID	Your user id obtained from the authentication step.
access_token	Your account access token.
includeExecutions	If true, include executions in the result.

HTTP Headers

Name	Value
x-qx-client-application	Defaults to qiskit-api-py

Response Sample

Listing 3-6 shows a sample response from this call.

Listing 3-6. Experiment List Response

```
{
  "total": 17,
  "count": 17,
  "codes": [{
    "type": "Algorithm",
    "active": true,
    "versionId": 1,
    "idCode": "...",
    "name": "3Q GHZ State YXY-Measurement 1",
    "jsonQASM": {
      ...
    },
    "qasm": "",
    "codeType": "QASM2",
    "creationDate": "2018-04-14T19:09:51.382Z",
    "deleted": false,
```

```
  "orderDate": 1523733740504,
  "userDeleted": false,
  "displayUrls": {
    "png": "URL"
  },
  "isPublic": false,
  "id": "...",
  "userId": "..."
}]}
```

Run Experiment

This call runs an experiment remotely in IBM Q Experience.

- **HTTP Method:** POST

- **URL:** https://quantumexperience.ng.bluemix.net/api/codes/
  execute?access_token=ACCESS-TOKEN&shots=SHOTS&deviceRun
  Type=RUN-TYPE

Request Parameters

Name	Value
shots	The number of shots to perform. The higher the number, the better the accuracy of your results will be. Note that this number depletes your credits at a rate of 3 credits per 1024 shots. Space is at a premium in the quantum world.
access_token	Your account access token.
deviceRunType	The device where to run the experiment. This could be • A real device name such as ibmqx2 and ibmqx3 for real processors. • For simulators: simulator or sim_trivial_2.
seed (optional)	An optional random number required only for simulators.

HTTP Headers

Name	Value
x-qx-client-application	Defaults to qiskit-api-py
Content-Type	application/json

Payload Format

The request body is a JSON document that describes the experiment as shown in the following snippet:

```
{
  "name": "Experiment NAME",
  "codeType": "QASM2",
  "qasm": "CODE"
}
```

Response Sample

This is arguably the most important call of the API. As an exercise, let's take one of the Bell states from the previous section and run it in both the simulator and real device using the REST API.

Listing 3-7. Bell State XW Measurement

```
IBMQASM 2.0;
include "qelib1.inc";

qreg q[2];
creg c[2];

h q[0];
cx q[0],q[1];
h q[0];
s q[1];
h q[1];
```

```
t q[1];
h q[1];
measure q[0] -> c[0];
measure q[1] -> c[1];
```

Listing 3-7 shows the assembly code from one of the Bell state (XW) experiments performed with the web console in the previous section. Take this code and create a JSON payload of the form: {"name": "NAME", "codeType": "QASM2", "qasm": "ONE-LINE-QASM"}. Note that we must give it an experiment name and the QASM code must be formatted in a single line including line feeds (\n); thus the final payload becomes

```
{"name": "REST Bell State XW", "codeType": "QASM2", "qasm": "IBMQASM 2.0;\
ninclude \"qelib1.inc\";\nqreg q[2];\ncreg c[2];\nh q[0];\ncx q[0],q[1];\nh
q[0];\ns q[1];\nh q[1];\nt q[1];\nh q[1];\nmeasure q[0] -> c[0];\nmeasure
q[1] -> c[1];"}.
```

Now we are ready to submit our experiment via REST. Don't forget that you must authenticate first to obtain an access token.

Tip   REST clients are available for most if not all current browsers. Install your favorite browser REST client, and create an authentication request as described in the "Authentication" section. Save it and keep it handy to obtain your access token.

I will use Chrome's YARC (Yet Another REST Client) to submit the payload to the simulator first and then to the real device (see Figure 3-8).

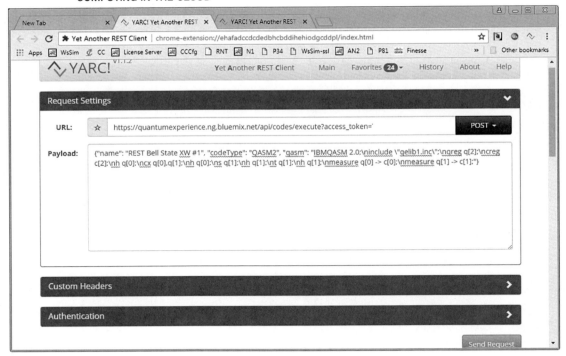

Figure 3-8.   *Chrome's YARC REST client with payload for Bell state XW
experiment*

Submit to the Simulator

To submit to the simulator, use the following request parameters:

```
access_token=ACESS_TOKEN&shots=1&deviceRunType=simulator
```

Verify that the response code is 200 (OK) and look at the response output. Verify that
the experiment has been recorded in the Q Experience console (see Figure 3-9).

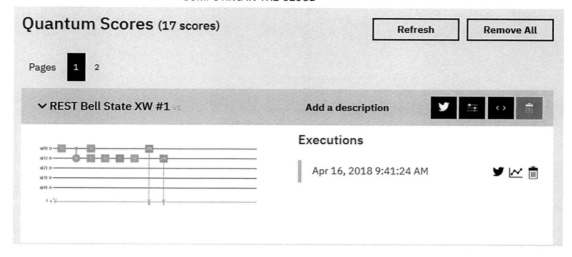

Figure 3-9. *Web console showing the Bell state XW experiment submitted via REST*

Submit to a Real Device

To submit to the real device (ibmqx4 in this case), change the request parameters to

```
access_token=ACESS_TOKEN &shots=1&deviceRunType=ibmqx4
```

Note   Real quantum devices may be offline for maintenance or whatever reason.
If that is the case, the submission will fail with a 400 (bad request) HTTP response.
Make sure the device is online before submitting to a real quantum device!

If all goes well, the job should be put in the execution queue and recorded in the web
console. Listing 3-8 shows the result from a submission to a real device with a status of
PENDING_IN_QUEUE.

Listing 3-8. Simplified HTTP Response from the Bell State XW Experiment
Submitted via REST

```
{
  "startDate": "2018-04-16T13:05:43.440Z",
  "modificationDate": 1523883943441,
  "typeCredits": "plan",
```

```
"status": {
  "id": "WORKING_IN_PROGRESS"
},
"deviceRunType": "real",
"ip": {
  "ip": "...",
  "city": "Raleigh",
  "country": "United States",
  "continent": "North America"
},
"shots": 1,
"paramsCustomize": {},
"deleted": false,
"userDeleted": false,
"id": "...",
"codeId": "...",
"userId": "...",
"infoQueue": {
  "status": "PENDING_IN_QUEUE",
  "position": 21,
  "estimatedTimeInQueue": 735
},
"code": {
  "type": "Algorithm",
  "active": true,
  "versionId": 1,
  "idCode": "...",
  "name": "REST Bell State XW #1",
  "jsonQASM": {
    ...
    "numberGates": 7,
    "hasMeasures": true,
    "numberColumns": 11,
    "include": "include \"qelib1.inc\";"
  },
```

```
    "qasm": "...",
    "codeType": "QASM2",
    "creationDate": "2018-04-16T13:05:42.547Z",
    "deleted": false,
    "orderDate": 1523883943351,
    "userDeleted": false,
    "isPublic": false,
    "id": "...",
    "userId": "..."
  }
}
```

At this point, you have succeeded submitting your first experiment via REST. Try playing by increasing the number of shots of your experiment to achieve greater accuracy.

Run a Job

This call is very similar to the previous *Run Experiment*; however it features two end points:

- **End point 1:** For regular users of the IBM Q Experience.

- **End point 2:** For corporate customers. It requires a hub, group, and project ids.

Corporate customers have premium access as well as access to the powerful 20-qubit processors and perhaps the rumored 50-qubit chip coming by the end of 2018.

- **HTTP Method:** POST

- **URL 1:** (5, 16 qubits) `https://quantumexperience.ng.bluemix.net/api/Jobs?access_token=ACCESS-TOKEN`

- **URL 2:** (20+ qubits corporate) `https://quantumexperience.ng.bluemix.net/api/Network/HUB/Groups/GROUP/Projects/PROJECT/jobs?access_token=ACCESS-TOKEN`

Request Parameters

Name	Value
access_token	Your account access token

HTTP Headers

Name	Value
x-qx-client-application	Defaults to qiskit-api-py
Content-Type	application/json

Payload Format

The format of the payload embeds all execution parameters: backend name, shots, and code in a single JSON document as shown in the following snippet:

```
{
  "backend": {
    "name": "simulator"
  },
  "shots": 1,
  "qasms": [{
    "qasm": "qams"
  }, ...]
}
```

Tip   Experiments submitted through the Run Job end point are not recorded in the *Scores* section of the Composer but put in an execution queue for processing.

On the other hand, a submission via the Run Experiment end point will record an entry in the Composer. Also, note that any experiment submitted to the simulator will return results immediately. Experiments submitted to a real quantum device will always enter the execution queue in a PENDING state. On completion, a notification email will be sent to the user. Let's send a quick job to the real device ibmqx4. Paste the following end point into your REST client:

```
https://quantumexperience.ng.bluemix.net/api/Jobs?access_token=access_token
```

Set the HTTP Method to POST, access token, and the headers as described in the previous section. Use the following JSON payload:

```
{
"qasms": [{
 "qasm": "\n\ninclude \"qelib1.inc\";\nqreg q[5];\ncreg c[5];\nu2(-4*pi/
 3,2*pi) q[0];\nu2(-3*pi/2,2*pi) q[0];\nu3(-pi,0,-pi) q[0];\nu3(-pi,0,
 -pi/2) q[0];\nu2(pi,-pi/2) q[0];\nu3(-pi,0,-pi/2) q[0];\nmeasure q -> c;\n" }],
 "shots": 1024,
 "backend": {
  "name": "ibmqx4"
 },
 "maxCredits": 3
}
```

The preceding payload submits a random experiment to the real device ibmqx4. Make sure it is online before submission (or use simulator instead). Also make sure the QASM code is in a single line including line feeds (\n). Note that double quotes must be escaped. If the submission fails, it probably means that the device is offline or your QASM payload is invalid. Double- and triple-check to make sure they are correct. The response I got tells me that my job is RUNNING:

```
{
  "qasms": [
    {
      "qasm": "...",
      "status": "WORKING_IN_PROGRESS",
      "executionId": "5ba6955fd867ef0046615172"
    }
  ],
  "shots": 1024,
  "backend": {
```

```
    "id": "5ae875670f020500393162b3",
    "name": "ibmqx4"
  },
  "status": "RUNNING",
  "maxCredits": 3,
  "usedCredits": 3,
  "creationDate": "2018-09-22T19:17:51.448Z",
  "id": "5ba6955fd867ef0046615171",
  "userId": "5ae875060f0205003931559a",
  "infoQueue": {
    "status": "PENDING_IN_QUEUE",
    "position": 11
  }
}
```

Note that my job won't show up in the Composer; however I will get an email with a link to the results.

Get the API Version

It returns the version of the Q Experience REST API.

- **HTTP Method:** GET

- **URL:** https://quantumexperience.ng.bluemix.net/api/
  version?access_token=ACCESS-TOKEN

Request Parameters

Name	Value
access_token	Your account access token

HTTP Headers

Name	Value
x-qx-client-application	Defaults to qiskit-api-py

Response Format

It returns a string with the version of the API, by the time of this writing 6.4.8.

We have peaked inside the Python IBMQuantumExperience REST API to see what goes on behind the scenes. As an exercise let's build a custom client for Node JS. Here is how.

A Node JS Client for the IBMQuantumExperience

This section presents a simple exercise to mimic one of the components of the Python SDK also known as the QISKit (Quantum Information Software Kit). In the next chapter, we will dive into the Python SDK in more detail, but we'll start by saying that the SDK is built on top of two basic libraries:

- **IBMQuantumExperience:** This is a REST client implementation in Python which I peeked into to present the REST API from the previous section. This library is not well documented. It makes sense as it is meant to be a modularized library that may change in the future.

- **QISKit SDK:** This is the main entry point to all your quantum programs. It packs gate logic; assembly translation; simulators, a local python simulator and a fast C++ simulator; and much more. This library also invokes IBMQuantumExperience for all interactions via REST to the IBM Q Experience platform.

Python is a great language, but Node JS is all the craze right now in the data center; thus this section presents a simple implementation of the REST API for Node JS. Here are some reasons why this would be a useful library:

- Node JS is a powerhouse for network I/O asynchronous calls. It is fast, and it is the perfect platform for REST clients.

- Python is a good language, with a plethora of amazing numerical, math, chart libraries, but has some idiosyncrasies that some may find unattractive. For example, indentation is relevant in Python (you cannot mix TABS and spaces – there are no braces to differentiate blocks). This drove me crazy at the beginning and took me a while to figure out as almost all computer languages use braces to

127

differentiate logic blocks. No such thing in Python, you must use
spaces or TABS and you cannot mix them. I don't like this. I think it
is bad design because if you make a mistake, you don't know where a
logic block ends. That's what braces are for.

- Variety is always good: after all my troubles with Python, I thought
  this library could be the foundation for a QISKit clone for Node JS.

So let's get started.

Build a Node Module for IBMQuantumExperience

I have tried to keep names as close as possible to the Python version. To create a Node
JS module, create a folder named IBMQuantumExperience, and initialize Node JS as
shown in the following set of commands:

```
$ mkdir IBMQuantumExperience
$ cd IBMQuantumExperience
$ npm init
```

Tip   Linux users: I have used Windows to write this section; furthermore I assume
that you have installed Node JS, are familiar with Node modules, and can code
in Javascript. All in all, you can get Node installers for all platforms at `https://
nodejs.org/en/`. Also note that Linux complains about uppercase names for
package names. Save yourself a headache and use Windows to go thru this
section.

Node provides a package manager called npm (which is pretty much the same thing
as Python's pip) which can be used to initialize your module. The third command will
create two files in the current folder:

- **index.js:** This is your module code. Note for Linux users, this file may
  not be created. If so create it manually.

- **Package.json:** This is the module descriptor with information such
  as name, version, author, dependencies, and more.

Create a folder test in the same location to contain unit tests and install the powerful
Node rest client *request*:

```
$mkdir test
$ npm install request
```

The second command installs the popular HTTP request package and all its
dependencies in the current folder. Now we are ready to start implementing the REST
API from the previous section. Open index.js in your favorite editor and let's get started.

Export API Methods

To expose an API via Node, use the module.exports library as shown in Listing 3-9.
Note that this is a partial implementation of the library, as the pieces are assembled
throughout the following sections. Nevertheless, a full implementation is available from
the book source under Workspace\Ch03\IBMQuantumExperience. That being said, you
should be able to paste all these listings into index.js. Again, I am assuming here that you
are familiar on how modules are written in Node JS.

Listing 3-9. Expose Public API Methods via Node

```
const log = require('./log'); // A simple custom log library (see the debug
section)
const request = require('request');

//...
module.exports =  {
  init: function (cfg) {
    _config = cfg;
    var debug = _config.debug ? _config.debug : false;
    log.init (debug);
    return loginWithToken ();
  },

  getCalibration  : calibration,
  getBackends     : backends,
  getParameters   : parameters,
  runExperiment   : experiment
```

```
  // Left as exercise  getJobs       : jobs,
  // Left as exercise  getMyCredits  : credits
}
```

In Listing 3-9 we import an external library using the require keyword:

- Line 2 imports the *request* HTTP client library to interact with Q
  Experience.

- Line 5 declares the public methods to be exposed by this module:

  - **init:** This method authenticates against the Q Experience
    platform as described in the previous section "Remote Access via
    the REST API."

  - **getCalibration:** It returns the platform's calibration parameters
    for a given device.

  - **getBackends:** It returns a list of quantum devices (and
    simulators) available for use.

  - **getParameters:** It returns the device parameters as described
    under the Devices section of the Composer web console.

  - **runExperiment:** It runs an experiment remotely in the simulator
    or real quantum device.

  - **getJobs:** It returns a list of current jobs in the experiment
    execution queue.

  - **getMyCredits:** It returns the user's execution credit and other
    useful information.

Authenticate with a Token

Before authentication, the library is initialized the same way as in Python, with a
configuration JSON object containing the Platform URL, API token, and more. For
example, this is how we would test the get backend information REST API call (taken
from test.js):

```
// require the `index.js` file from the same directory.
const qx = require('index.js');
```

```
// Put your API token here
var config = { APItoken: 'YOUR_API_TOKEN'
  , debug     : true
  , 'url'     : 'https://quantumexperience.ng.bluemix.net/api'
  , 'hub'     : 'MY_HUB'
  , 'group'   : 'MY_GROUP'
  , 'project' : 'MY_PROJECT'
}
// Get backend information
async function testBackends() {
  await qx.init(config);
  var result = await qx.getBackends();
  console.log("---- BACKENDS ----\n" + JSON.stringify(result) + "\n-----" );
}
```

Remember that hub, group, and project are parameters for corporate customers only; thus they are not used in this implementation; however support for them can be easily added. Once initialized, simply submit a POST request to log in with a token as described by the REST API (see Listing 3-10). Note that this code, as well as all the listings throughout these sections, goes in index.js.

Listing 3-10. Token Authentication via REST

```
function loginWithToken () {
  let options = {
    url: _config.url + '/users/loginWithToken',
    form: {'apiToken': _config.APItoken}
  };

  return new Promise(function(resolve, reject) {
    // Do async job
    // {"id":"Access tok","ttl":1209600,"created":"2018-04-
       17T23:30:21.089Z","userId":"userid"}
    request.post(options, function(err, res, body) {
      if (err) {
        reject(err);
      }
```

```
    else {
        var json      = JSON.parse(body);
        _accessToken  = json.id;
        _userId       = json.userId;
        log.debug("Got User:" + _userId + " Tok:" +  _accessToken);
        resolve(JSON.parse(body));
      }
    });
  })
}
```

In Listing 3-10

- The `request.post` system call is used to send an HTTP POST to
  the end point `https://quantumexperience.ng.bluemix.net/api/`
  `users/loginWithToken` using the JSON payload `{'apiToken':`
  `'YOUR_TOKEN'}`. As described by the REST API, this call returns a new
  JSON document: `{"id":"TOKEN","ttl":1209600,"created":`
  `"DATE","userId":"USERID"}`. This document is parsed and the
  access token (id) and user id (userId) saved for later use.

- Note that because all network I/O in Node is asynchronous, all
  methods return a Promise. This is basically an asynchronous task that
  encapsulates the difficulty of having to wait for the task to complete
  before reading results. Thus if the HTTP request call succeeds, the
  resolve callback from the Promise will fire with the HTTP response
  data; else the Promise reject callback will be invoked.

Tip   Promises are a compelling alternative to callbacks for asynchronous code.
Nevertheless they can be confusing some times. All in all, Promises are becoming
the de facto standard for asynchronous programming in Javascript.

Nevertheless if you find the Promise handling code convoluted, there is an even
easier way to deal with this, and it is shown in the next section where we implement a
method to fetch the list of backends.

List Backends

Listing 3-11 shows a Node request to fetch backends from Q Experience.

- It sends an HTTP GET request to `https://quantumexperience.ng.bluemix.net/api/Backends?access_token=TOKEN`.

- It returns a Promise which can be called within any asynchronous function by using the new `async`/`await` feature in Javascript.

Listing 3-11. Get Backend List via Node

```
const _defaultHdrs = {
    'x-qx-client-application': _userAgent
};
function backends () {
  let options = {
    url: _config.url + '/Backends?access_token=' + _accessToken,
    headers: _defaultHdrs
  };
  return new Promise(function(resolve, reject) {
    // Do async job
    request.get(options, function(err, res, body) {
      if (err) {
        reject(err);
      }
      else {
        resolve(JSON.parse(body));
      }
    });
  })
}
```

To test the preceding method, we can use the `async/await` feature from Node.js >=7.6 as shown in the following snippet:

```
async function testBackends() {
  await qx.init(config);
  var result = await qx.getBackends();
  console.log("---- BACKENDS ----\n" + JSON.stringify(result) + "\n-----" );
}
```

Tip   An *async* function can contain an *await* expression that pauses the execution of the async function and waits for the passed Promise's resolution and then resumes the async function's execution and returns the resolved value.

List Calibration Parameters

Listing 3-12 shows how to get calibration and hardware parameters for a specific backend in IBM Q Experience:

- Fetch calibration information by sending a GET request to `https://quantumexperience.ng.bluemix.net/api/Backends/NAME/calibration?access_token=TOKEN` where NAME is the backend you wish to query and TOKEN is the access token obtained from the authentication step.

- Fetch backend parameters by sending a similar GET request to `https://quantumexperience.ng.bluemix.net/api/Backends/NAME/parameters?access_token=TOKEN`.

- The response format for both requests is described in the Remote Access via the REST API section of this chapter.

Listing 3-12. Get Device Calibration and Parameter Data

```
function calibration (name) {
  let options = {
    url: _config.url + '/Backends/' + name +'/calibration?access_token=' +
    _accessToken,
    headers: _defaultHdrs
  };
  return new Promise(function(resolve, reject) {
    request.get(options, function(err, res, body) {
      if (err) {
        reject(err);
      }
      else {
        resolve(JSON.parse(body));
      }
    });
  })
}
function parameters (name) {
  let options = {
    url: _config.url + '/Backends/' + name +'/parameters?access_token=' +
    _accessToken,
    headers: _defaultHdrs
  };
  return new Promise(function(resolve, reject) {
    request.get(options, function(err, res, body) {
      if (err) {
        reject(err);
      }
      else {
        resolve(JSON.parse(body));
      }
    });
  })
}
```

To test the code, create an async function and use the await keyword to get the response from the asynchronous task as shown in the following snippet:

```
async function testCalibration() {
  await qx.init(config);
  var result1 = await qx.getCalibration('ibmqx4');
  var result2 = await qx.getParameters('ibmqx4');
  console.log(JSON.stringify(result1) );
  console.log(JSON.stringify(result1) );
}
```

For the final step, let's see how an experiment can be run.

Run the Experiment

This is the most important call of the API, and once executed, the experiment should be recorded under the scores section of your IBM Q Experience web console (see Listing 3-13). To submit an experiment programmatically, send an HTTP POST to the /codes/execute end point with the JSON payload:

```
{'name': name, "codeType": "QASM2", "qasm":  "YOUR_QASM_CODE"}
```

- Remember that the assembly code must be formatted in a single line with line feeds (\n) to separate instructions. For example, the following code declares 5 qubits and 5 classical registers: "\n\ ninclude \"qelib1.inc\";\nqreg q[5];\ncreg c[5];\n".

- The name parameter defines the experiment name to be recorded in the web console.

- The shots parameter is the number of shots executed by the quantum processor.

- The device parameter can be simulator (for the remote simulator) or a real quantum device name such as ibmqx4.

Tip   If you run an experiment in a real device, it will enter an execution queue
for future processing. You will receive an email on completion. On the other hand,
if you run the experiment in the remote simulator, the results will be returned
synchronously.

Listing 3-13. Run an Experiment

```
const _userAgent = 'qiskit-api-py'; // A global

function experiment (name, qasm, shots, device) {
  let options = {
    url: _config.url + '/codes/execute?access_token=' + _accessToken
      + '&shots=' + shots + '&deviceRunType=' + device,
    headers: {'Content-Type': 'application/json', 'x-qx-client-
    application': _userAgent} ,
    form: {'name': name, "codeType": "QASM2", "qasm": qasm}
  };
  return new Promise(function(resolve, reject) {
    request.post(options, function(err, res, body) {
      if (err) {
        reject(err);
      }
      else {
        resolve(JSON.parse(body));
      }
    });
  })
}
```

Paste Listing 3-13 into index.js, use the following snippet to run an experiment in the real quantum device ibmqx4, then verify the experiment has been recorded in the web console, and finally wait for a notification email.

```
async function testExperiment () {
  await qx.init(config);
  var name    = "REST Experiment from Node JS #1"
  var qasm    = "\n\ninclude \"qelib1.inc\";\nqreg q[5];\ncreg c[5];\nu2(-
  4*pi/3,2*pi) q[0];\nu2(-3*pi/2,2*pi) q[0];\nu3(-pi,0,-pi) q[0];\nu3(-pi,0,-
  pi/2) q[0];\nu2(pi,-pi/2) q[0];\nu3(-pi,0,-pi/2) q[0];\nmeasure q -> c;\n";
  var shots   = 1;
  var device  = "ibmqx4";
  var result  = await qx.runExperiment(name, qasm, shots, device);
  console.log("---- EXPERIMENT " + name + " ----\n" + JSON.
stringify(result) + "\n-----" )
}
```

Tip   The code for the IBMQuantumExperience Node module is included in the book source under Workspace\Ch03\IBMQuantumExperience. The project has a test script under test/tests.js. Edit this file, add your API token, and execute it from IBMQuantumExperience with the command: node `test/tests.js`.

Debugging and Testing

For simple debugging I have created the submodule log.js (at the same level as index.js) and used the quintessential console object to display information into the console as shown in the following snippet:

```
var _debug = false;

function LOGD( tag, txt ) {
  if ( _debug ) {
    console.log('[DBG-QX] ' + tag + ' '  + (txt ? txt : "));
  }
}
```

```
function LOGE( tag, txt ) {
  console.error('[ERR-QX] ' + tag + ' ' + (txt ? txt : "));
}
function init (debug) {
  _debug = debug;
}

exports.init = init;
exports.debug = LOGD;
exports.error = LOGE;
```

The main module (index.js) uses this submodule to display debug messages in the console. Finally, to test the package, edit test/tests.js and paste the test snippets described throughout these sections as shown in the following partial listing from tests.js:

```
// test/tests.js require the `index.js` file from the same directory.
const qx = require('../');

// Put your API token here
var config = { APItoken: 'API-TOKEN'
  , debug: true
  , 'url': 'https://quantumexperience.ng.bluemix.net/api'
  , 'hub': 'MY_HUB'
  , 'group': 'MY_GROUP'
  , 'project': 'MY_PROJECT'
};

async function testBackends() {
 await qx.init(config);
 var result = await qx.getBackends();
 console.log("---- BACKENDS ----\n" + JSON.stringify(result) + "\n-----" );
}

async function testJobs () {
 await qx.init(config);
 var filter = '{"limit":2}';
 var jobs = await qx.getJobs(filter);
 console.log ("---- JOBS----\n" + JSON.stringify(jobs) + "\n----");
}
```

139

```
// Paste all test snippets here...
// ....

try {
 testBackends();
 testJobs ();
 // more tests here...
}
catch (e){
 console.error(e);
}
```

To run the test, execute `node test\tests.js` within the IBMQuantumExperience folder. Note that I have left out two methods: getJobs and getMycredits as an exercise. With this foundation, you should be able to easily implement and test them.

Share with the World: Publish Your Module

If you wish to share your work with the world, you can publish your module to the npm registry. For this you must create a user account at `www.npmjs.com/` or manually using the commands

```
npm adduser
npm publish
```

Make sure you document your code by adding a markdown document (readme.md) to the root folder. After publishing, navigate to `https://npmjs.com/package/<package>` and check out your live module. Now others should be able to install it with

```
npm install IBMQuantumExperience
```

Node JS developers can now submit experiments to the Q Experience with code like this:

```
const qx = require('IBMQuantumExperience');
...
async function sendExperiment () {
  var config = { APItoken: 'API-TOKEN'
    , 'url': 'https://quantumexperience.ng.bluemix.net/api', 'debug': false};
```

```
await qx.init(config);
var name     = "REST Experiment from Node JS #1"
var qasm     =  "MY_QASM";
var device   = "ibmqx4";
var result   = await qx.runExperiment(name, qasm, 1024, device);
}
```

In this chapter you have taken the first step in your new career as a quantum programmer. IBM has created an amazing cloud platform to learn about these incredible machines. We should thank the good folks at IBM for making this platform freely accessible to the masses. For now, quantum computers are experimental machines, so don't expect to get one at the local hardware store. Nevertheless soon they will take over the data center, so now is the time to learn how to program them.

CHAPTER 4

QISKit, Awesome SDK for Quantum Programming in Python

In this chapter you will get started with the QISKit, the best SDK out there for quantum programming. You will learn how easy it is to install the SDK in your local system. But before writing your first quantum program, it is always helpful to understand what quantum computation is and how it differs from classical computation. For this purpose, a very basic explanation of qubit states and quantum gates is presented using linear algebra. This section also shows how quantum computation can mirror its classical counterpart and furthermore find shortcuts to get results faster. Next, the chapter walks through the anatomy of a quantum program including system calls, circuit compilation formats, quantum assembly, and more.

QISKit packs a set of helpful simulators to execute your programs locally or remotely, but it also allows you to run in the real thing. Step by step, you will learn how to run your quantum programs in a real device provided by the awesome IBM Q Experience cloud platform. So start your desktop and let's get to it.

Installing the QISKit

QISKit is the Quantum Information Software Kit, the de facto SDK for quantum programming in the cloud. It is written in Python, a powerful scripting language for scientific computing. My background has been mostly in business so I haven't written much Python code over the years, so let's see how the SDK can be installed both in Linux CentOS 6–7 and in Windows 64. We'll begin with the easiest (Windows) and then jump to the trickiest (CentOS).

© Vladimir Silva 2018
V. Silva, *Practical Quantum Computing for Developers*, https://doi.org/10.1007/978-1-4842-4218-6_4

Setting Up in Windows

QISKit requires Python 3.5 or later. If you have a Windows system, chances are that you don't have Python installed. If so, you can get the installers from the Python.org web site. Download the installer, run it, and verify your installation by running the following from the command window:

```
C:\>Python -V
Python 2.7.6
```

I have good old Python 2.7, yet you can have multiple versions of Python installed at the same time. In my case I downloaded the embeddable zip file and deployed to C:\Python36-64 so in my case:

```
C:\>C:\Python36-64\Python.exe -V
Python 3.6.4
```

Python features an amazing package manager called pip (preferred installer program) which makes installing modules very easy. Thus, to install QISKit, simply type at the console:

```
C:\>pip install qiskit
```

Your screen output should look similar to Listing 4-1. Make sure there are no error messages.

Listing 4-1. QISKit Installation in Windows 64 Bit

```
Collecting qiskit
  Using cached qiskit-0.4.11.tar.gz
Collecting IBMQuantumExperience>=1.8.28 (from qiskit)
  Using cached IBMQuantumExperience-1.9.0-py3-none-any.whl
Collecting matplotlib<2.2,>=2.1 (from qiskit)
  Using cached matplotlib-2.1.2.tar.gz
Collecting networkx<2.1,>=2.0 (from qiskit)
  Downloading networkx-2.0.zip (1.5MB)
     100% |████████████████████████████████| 1.6MB 400kB/s
```

```
Collecting numpy<1.15,>=1.13 (from qiskit)
  Downloading numpy-1.14.2-cp36-cp36m-manylinux1_i686.whl (8.7MB)
    100% |████████████████████████████████| 8.7MB 105kB/s
...
  Running setup.py install for pycparser ... done
  Running setup.py install for matplotlib ... done
  Running setup.py install for networkx ... done
  Running setup.py install for ply ... done
  Running setup.py install for mpmath ... done
  Running setup.py install for sympy ... done
  Running setup.py install for qiskit ... done
Successfully installed IBMQuantumExperience-1.9.0 qiskit-0.4.11
requests-2.18.4 ...
```

This is it; you have taken the first step in this journey as a quantum programmer. For the Linux user, let's set things up in CentOS 6 or 7.

Setting Up in Linux CentOS

Things are a bit trickier to set up in CentOS 6 or 7. This is due to the fact that CentOS focuses mainly in stability than bleeding edge software. Thus CentOS comes with Python 2.7 out of the box; furthermore the official distribution does not provide packages for Python 3.5. This doesn't mean however that Python 3.5 cannot be installed. Let's see how.

Tip   The instructions in this section should work for any Linux flavor based on the Red Hat base such as RHEL 6–7, CentOS 6–7, and Fedora Core.

Step 1: Prepare Your System

First make sure that yum (the Linux Update Manager) is up to date by running the command:

```
$ sudo yum -y update
```

Next, install yum-utils, a collection of utilities and plugins that extend and supplement yum:

```
$ sudo yum -y install yum-utils
```

Install the CentOS development tools. These include compilers and libraries to allow for building and compiling many types of software:

```
$ sudo yum -y groupinstall development
```

Now, let's install Python 3. Note that we'll run multiple versions of Python: the official, 2.7, and 3.6 for development.

Step 2: Install Python 3

To break out of the chains of the default CentOS distribution, we can use a community project called *Inline with Upstream Stable* (IUS). This is a set of the latest development libraries for OSes that don't provide them such as CentOS. Let's install IUS in it through yum:

```
$ sudo yum -y install https://centos7.iuscommunity.org/ius-release.rpm
(CentOS7)
$ sudo yum -y install https://centos6.iuscommunity.org/ius-release.rpm
(CentOS6)
```

Once IUS is finished installing, we can install the most recent version of Python (3.6):

```
$ sudo yum -y install python36u
```

Check to make sure that the installation is correct:

```
$ python3.6 -V
Python 3.6.4
```

Now, let's install pip and verify:

```
$ sudo yum -y install python36u-pip
$ pip3.6 -V
```

Finally, we will need to install the IUS package python36u-devel, which provides useful Python development libraries:

```
$ sudo yum -y install python36u-devel
```

Step 3: Don't Disturb Others – Set Up a Virtual Environment

This step is useful only if you have a multiuser system running multiple versions of Python and don't want to disturb other users. For example, to create a virtual environment in your home folder:

```
$ mkdir $HOME/qiskit
$ cd $HOME/qiskit
$ python3.6 -m venv qiskit
```

The preceding command sequence creates a folder called qiskit in the user's home to contain all your quantum programs. Inside this folder, a virtual Python 3.6 environment called qiskit is also created. To activate the environment, run the command:

```
$ source qiskit/bin/activate
(qiskit) [centos@localhost qiskit]$
```

Within the virtual environment, you can use the command python instead of python3.6 and pip instead of pip3.6 if you prefer:

```
$ python -V
Python 3.6.4
```

Tip   If you don't activate your virtual environment, then you must use python3.6 and pip3.6 instead of python and pip.

Step 4: Install QISKit

Activate your virtual environment and install QISKit with the command:

```
$ pip install qiskit
```

Listing 4-2 shows the standard output of the preceding command.

Listing 4-2. QISKit Installation in CentOS 6

```
Collecting qiskit
  Downloading qiskit-0.5.7.tar.gz (4.5MB)
    100% |█████████████████████████████████████████| 4.5MB 183kB/s
Collecting IBMQuantumExperience>=1.8.28 (from qiskit)
  Downloading IBMQuantumExperience-1.9.0-py3-none-any.whl
Collecting matplotlib<2.2,>=2.1 (from qiskit)
  Downloading matplotlib-2.1.2.tar.gz (36.2MB)
    100% |█████████████████████████████████████████| 36.2MB 18kB/s
    Complete output from command python setup.py egg_info:
    ====================================================================
    Edit setup.cfg to change the build options

    BUILDING MATPLOTLIB
                matplotlib: yes [2.1.2]
                    python: yes [3.6.4 (default, Dec 19 2017, 14:48:15)  [GCC
                            4.4.7 20120313 (Red Hat 4.4.7-18)]]
                  platform: yes [linux]
...
Installing collected packages:  IBMQuantumExperience,
numpy, python-dateutil, pytz, cycler, pyparsing, matplotlib,
decorator, networkx, ply, scipy, mpmath, sympy, pillow, qiskit
  Running setup.py install for pycparser ... done
  Running setup.py install for matplotlib ... done
  Running setup.py install for networkx ... done
  Running setup.py install for ply ... done
  Running setup.py install for mpmath ... done
  Running setup.py install for sympy ... done
  Running setup.py install for qiskit ... done
Successfully installed IBMQuantumExperience-1.9.0  qiskit-0.4.11
requests-2.18.4 requests-ntlm-1.1.0 scipy-1.0.1 six-1.11.0 sympy-1.1.1 urllib3-1.22

(qiskit) [centos@localhost qiskit]$
```

Tip   Under a virtual environment, Python packages will be installed in the environment's home lib/python3.6/site-packages instead of the system's path as shown in Figure 4-1.

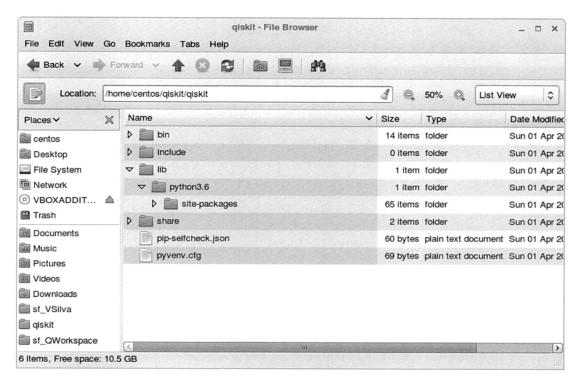

Figure 4-1. *Python virtual environment folder layout*

We are now ready to start writing quantum code. Let's see how.

Qubit 101: It's Just Basic Algebra

Before we start writing quantum programs, we need to refresh some fundamental mathematics to understand what goes on behind the scenes. In the previous chapter, you learned how a qubit is represented by the Bloch sphere: a geometrical representation of the pure state of a two-level quantum mechanical system (qubit). But perhaps a better way of understanding the basic model of the qubit and the effects of quantum gates is to use its algebraic representation. For this purpose you need to dust up some basic linear algebra concepts including

- *Linear vectors*: Simple vectors such as $\begin{bmatrix} 1 \\ 0 \end{bmatrix}$ which will be used to represent the basis states of the qubit.

- *Complex number*: A complex number is a number composed of a real and imaginary parts denoted by a + bi where $i = \sqrt{-1}$. Note that complex numbers cannot exist in our physical reality. The coefficients α, β of the super imposed state of a qubit $\psi = \alpha \,|\,0\rangle + \beta \,|\,1\rangle$ are complex numbers.

- *Complex conjugate*: A term that you will often hear when talking about quantum gates. To obtain a complex conjugate, simply flip the sign of the imaginary part; thus a + bi becomes a – bi and vice versa.

- *Matrix multiplication*: If A is an n × m matrix and B is an m × p matrix, their product AB is an n × p matrix, in which the m entries across a row of A are multiplied with the m entries down a column of B and summed to produce an entry of AB. Take the first row from the first matrix and multiply each element for the first column of the second matrix, which becomes the first element in the result matrix (see Figure 4-2); but don't panic, most of the matrices that we will be looking at are 2x2 consisting of 0 or 1 elements.

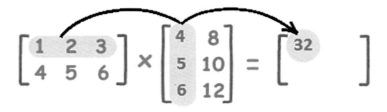

Figure 4-2. *Basic matrix multiplication operation*

Algebraic Representation of a Quantum Bit

In the classical model, the fundamental unit of information is the bit which is represented by a 0 or 1. The bit physically translates to the voltage flow through a transistor. In quantum computation, the fundamental unit is the quantum bit (qubit) which physically translates to manipulations on photons, electrons, or atoms. Algebraically, the qubit is represented by the ket notation.

Tip   Ket notation was introduced in 1939 by physicist Paul Dirac and is also known as the Dirac notation. The ket is typically represented as a column vector and written as $|\varphi\rangle$.

Dirac's Ket Notation

Using Dirac's notation, the basic quantum states of the qubit are represented by the vectors |0> and |1>. These are called the computational basis states.

Tip   The quantum state of a qubit is a vector in a two-dimensional complex vector space. Let's illustrate this with a simple graph.

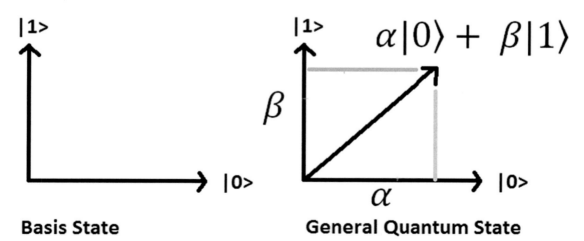

Basis State   **General Quantum State**

Figure 4-3. *Quantum states of the qubit*

Figure 4-3 shows the complex vector space used to represent the state of a qubit. On the left side, the so-called basis state is made up of two unit vectors in the Dirac notation for the states |0> and |1>. On the right side, a general quantum state is made up of a linear combination of the two. Thus, the basis states and general quantum states can be written as vectors:

$$|0\rangle = \begin{bmatrix} 1 \\ 0 \end{bmatrix}, |1\rangle = \begin{bmatrix} 0 \\ 1 \end{bmatrix}$$

$$\alpha|0\rangle + \beta|1\rangle$$

where α and β are amplitude coefficients of the unit vector. Note that a unit vector's amplitude must be 1; therefore α and β must obey the constraint $|\alpha|^2 + |\beta|^2 = 1$. This algebraic representation is the key to understanding the effect of a logic gate in the qubit as you will see later on.

So why is the state of a qubit represented as a vector in a seemly more complicated representation than its classical counterpart? Why vectors at all? The reason comes to that it allows for building a better model of computation as will be shown once we look at quantum gates and superposition of states. All in all, quantum mechanics is a theory that has evolved over many decades, and at the end of the day, a vector is a very simple mathematical object, easy to understand and manipulate. Probably the best tool for the job.

Superposition Is Just a Fancy Word

Superposition is defined by physicists as the property of atomic particles to exist in multiple states at the same time. If you find this concept difficult to grasp, then linear algebra can help.

Tip   Superposition is simply the linear combination of the |0> and |1> states. That is, $\alpha|0\rangle + \beta|1\rangle$ where the length of the state vector is 1 as shown in Figure 4-3.

Ket Notation Too Weird? Use Vectors Instead

If you like algebra and find the ket notation confusing, just use the familiar vector representation instead. Thus the superposition from the previous section can be written as

$$|\Psi\rangle = \alpha|0\rangle + \beta|1\rangle = \alpha\begin{bmatrix}1\\0\end{bmatrix} + \beta\begin{bmatrix}0\\1\end{bmatrix} = \begin{bmatrix}\alpha\\\beta\end{bmatrix}$$

Note that, because kets are vectors, they obey the same rules as vectors do, for example, multiplication by a scalar:

$$2\left(\alpha|0\rangle + \beta|1\rangle\right) = 2\begin{bmatrix}\alpha\\\beta\end{bmatrix} = \begin{bmatrix}2\alpha\\2\beta\end{bmatrix}$$

Changing the State of a Qubit with Quantum Gates

The purpose of quantum gates is to manipulate the state of a qubit to achieve a desired result. They are the basic building blocks of quantum computation just as classic logic gates are for the classical world. Some the quantum gates are the equivalent of their classical counter parts. Let's take a look.

NOT Gate (Pauli X)

This is the simplest gate and it acts in a single qubit. It is the quantum equivalent of the classical NOT gate, and just like its counterpart, it flips the state of the qubit. Thus

$$|0\rangle \rightarrow |1\rangle, |1\rangle \rightarrow |0\rangle$$

For a superposition, the X gate acts linearly, meaning it flips the corresponding state; thus |0> becomes |1> and |1> becomes |0>:

$$\alpha|0\rangle + \beta|1\rangle \rightarrow \alpha|1\rangle + \beta|0\rangle$$

153

In a quantum circuit, the NOT gate is represented by the X also known as Pauli X, named after Austrian physicist Wolfgang Pauli, one of the fathers of quantum mechanics.

q[0] $|0\rangle$ ——[X]——

The circuit starts with the basis state |0> for qubit 0, the state flows through the quantum wire until a manipulation is done in the state, and then the output continues through the wire.

There is another way of looking at the X gate in action; by using its matrix representation, we can see exactly how the state is flipped by using the Pauli matrix

$$X = \begin{bmatrix} 0 & 1 \\ 1 & 0 \end{bmatrix}$$

The state of the qubit is flipped by using the matrix representation of X and the vectors for $|0\rangle = \begin{bmatrix} 1 \\ 0 \end{bmatrix}$ and $|1\rangle = \begin{bmatrix} 0 \\ 1 \end{bmatrix}$; thus

$$X|0\rangle = \begin{bmatrix} 0 & 1 \\ 1 & 0 \end{bmatrix}\begin{bmatrix} 1 \\ 0 \end{bmatrix} = \begin{bmatrix} 0+0 \\ 1+0 \end{bmatrix} = \begin{bmatrix} 0 \\ 1 \end{bmatrix} = |1\rangle$$

$$X|1\rangle = \begin{bmatrix} 0 & 1 \\ 1 & 0 \end{bmatrix}\begin{bmatrix} 0 \\ 1 \end{bmatrix} = \begin{bmatrix} 0+1 \\ 0+0 \end{bmatrix} = \begin{bmatrix} 1 \\ 0 \end{bmatrix} = |0\rangle$$

There is an even simpler quantum circuit, the simplest of them all, and it is the quantum wire denoted by the Greek symbol (Psi) $|\psi\rangle$ _ _ _ _ _ _ _ _ _ _ $|\psi\rangle$ which describes the computational state over time. It may seem trivial, but physically this is the hardest thing to implement. Because of the atomic scale of the quantum wire (think photons, electrons, or single atoms), it is very fragile and prone to errors introduced by the environment.

Another interesting property of the X gate is that two NOT gates in a row give the identity matrix (I), a very important tool in linear transformations. Let's do the math:

q[0] $|0\rangle$ ——[X]——[X]——

$|\psi\rangle \rightarrow XX|\psi\rangle$

To understand the effects of the circuit, let's see what happens when we multiply two X matrices:

$$XX = \begin{bmatrix} 0 & 1 \\ 1 & 0 \end{bmatrix}\begin{bmatrix} 0 & 1 \\ 1 & 0 \end{bmatrix} = \begin{bmatrix} 0+1 & 0+0 \\ 0+0 & 1+0 \end{bmatrix} = \begin{bmatrix} 1 & 0 \\ 0 & 1 \end{bmatrix} = I$$

The X gate is the simplest example of a quantum logic gate, circuit, and computation. In the next section, we look at a truly quantum gate, Hadamard, and how it can trigger superpositions using circuits and algebra.

Truly Quantum: Superpositions with the Hadamard Gate

The effects of the Hadamard gate in the basis states are formally defined as

$$|0\rangle \to \frac{|0\rangle + |1\rangle}{\sqrt{2}}, |1\rangle \to \frac{|0\rangle - |1\rangle}{\sqrt{2}}$$

Furthermore, for a superposition state $\alpha|0\rangle + \beta|1\rangle$, the Hadamard maps to

$$\alpha|0\rangle + \beta|1\rangle \to \alpha\left(\frac{|0\rangle + |1\rangle}{\sqrt{2}}\right) + \beta\left(\frac{|0\rangle - |1\rangle}{\sqrt{2}}\right) = \frac{\alpha + \beta}{\sqrt{2}}|0\rangle + \frac{\alpha - \beta}{\sqrt{2}}|1\rangle$$

For the circuit and matrix presentation, the Hadamard acts on a single qubit.

q[0] $|0\rangle$ — [H] —

$$H = \frac{1}{\sqrt{2}}\begin{bmatrix} 1 & 1 \\ 1 & -1 \end{bmatrix}$$

Applying H to the basis states $|0\rangle = \begin{bmatrix} 1 \\ 0 \end{bmatrix}$ and $|1\rangle = \begin{bmatrix} 0 \\ 1 \end{bmatrix}$:

$$H|0\rangle = \frac{1}{\sqrt{2}}\begin{bmatrix} 1 & 1 \\ 1 & -1 \end{bmatrix}\begin{bmatrix} 1 \\ 0 \end{bmatrix} = \frac{1}{\sqrt{2}}\begin{bmatrix} 1 \\ 1 \end{bmatrix} = \frac{1}{\sqrt{2}}\left(\begin{bmatrix} 1 \\ 0 \end{bmatrix} + \begin{bmatrix} 0 \\ 1 \end{bmatrix}\right) = \frac{|0\rangle + |1\rangle}{\sqrt{2}}$$

$$H|1\rangle = \frac{1}{\sqrt{2}}\begin{bmatrix} 1 & 1 \\ 1 & -1 \end{bmatrix}\begin{bmatrix} 0 \\ 1 \end{bmatrix} = \frac{1}{\sqrt{2}}\begin{bmatrix} 1 \\ -1 \end{bmatrix} = \frac{1}{\sqrt{2}}\left(\begin{bmatrix} 1 \\ 0 \end{bmatrix} - \begin{bmatrix} 0 \\ 1 \end{bmatrix}\right) = \frac{|0\rangle - |1\rangle}{\sqrt{2}}$$

So what is the computational reason for the Hadamard gate? What does this buy us? Without getting too technical, the answer is that the Hadamard gate expands the range of states that are possible for a quantum circuit. This is important because the expansion of states creates the possibility of finding shortcuts and therefore doing computations faster. An analogy would be to a game of chess. For example, if your knight was allowed to move like a queen and knight at the same time (an expansion of states), this will tilt the game in your favor and allow you to checkmate faster. This is what Hadamard gives: more horsepower to your quantum machine.

Measurement of a Quantum State Is Trickier Than You Think

Imagine you have a lab in the basement of your home. You are given a qubit in state $|\psi\rangle = \alpha|0\rangle + \beta|1\rangle$ and a measurement apparatus and asked to calculate the α and β coefficients. That is, compute the quantum state. It may seem like a trivial task; however this is not possible. The principles of quantum mechanics state that the quantum state of a system is not directly observable. The best we can do is guess approximate information about α and β. This process is called measurement in the computational basis.

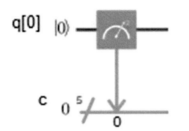

The outcome of a measurement on the quantum state $|\psi\rangle = \alpha|0\rangle + \beta|1\rangle$ gives the classical bits:

$\alpha|0\rangle + \beta|1\rangle \rightarrow 0$ *with probality* $|\alpha^2|$
$\alpha|0\rangle + \beta|1\rangle \rightarrow 1$ *with probality* $|\beta^2|$

Thus the measurement process spits the probabilities of the classical bits 0 and 1 equal to the absolute values of the coefficients α and β squared. Physically, the way to imagine this process taking place is by observing a physical photon, atom, or electron with a measurement apparatus. This is the reason why measurement is often regarded as a quantum gate.

Measurement disturbs the state of the quantum system giving a classical bit outcome. The important thing to remember is that, after the process, the coefficients α and β are destroyed. This means that we cannot store large amounts of information in a qubit. Imagine if we could measure the exact values for α and β, then by using complex numbers it would be possible in theory to store infinite amounts of classical information in the qubit state. By calculating the exact values of α and β, we could extract all that classical information. However this is not possible. Quantum mechanics forbids it.

One final point on measurement is the *normalization of the quantum state*: given a measurement in the computational basis $\alpha|0\rangle + \beta|1\rangle$, the probability of the classical bit 0 and 1 must add to 1. That is,

$$Probality(0) + Probality(1) = |\alpha^2| + |\beta^2| = 1$$

This means that the length of the quantum state vector must be 1 (normalized). This comes from the fact that measurement probabilities add to 1. In the next section, we'll talk about how single-qubit gates are generalized, what they are, and how they are used to build more complex circuits.

Generalized Single-Qubit Gates

So far we have seen two simple gates: X and H represented by the matrices:

$$X = \begin{bmatrix} 0 & 1 \\ 1 & 0 \end{bmatrix}, H = \frac{1}{\sqrt{2}} \begin{bmatrix} 1 & 1 \\ 1 & -1 \end{bmatrix}$$

Remember also that the superposition of the quantum state is expressed as the vector $|\Psi\rangle = \begin{bmatrix} \alpha \\ \beta \end{bmatrix}$. Then applying both gates to the quantum state can be generalized for any unitary matrix:

$$H \begin{bmatrix} \alpha \\ \beta \end{bmatrix}, X \begin{bmatrix} \alpha \\ \beta \end{bmatrix}, U \begin{bmatrix} \alpha \\ \beta \end{bmatrix} \text{ where } U = H, X$$

U is called the generalized single-qubit gate given the constraint that U must be unitary.

Tip   A matrix U is unitary if multiplied by its Hermitian transpose $U^\dagger$ it gives the identity matrix: $U^\dagger U = I$. The Hermitian transpose or conjugate transpose is denoted by a dagger (†) symbol $U^\dagger = (U^T)^*$, that is, the complex conjugate of the transposed.

The transpose of a matrix is a new matrix whose rows are the columns of the original. For example, if $A = \begin{bmatrix} a & b \\ c & d \end{bmatrix}$ then $A^T = \begin{bmatrix} a & c \\ b & d \end{bmatrix}$. Then, to obtain the Hermitian transpose $A^\dagger = \begin{bmatrix} a & c \\ b & d \end{bmatrix}^*$, take the complex conjugate of each entry. (The complex conjugate of a + bi, where a and b are real, is a – bi, that is, switch the sign of the imaginary part if any.)

Note that both gates H and X must be unitary. This can be easily verified by calculating $X^\dagger X = I$ and $H^\dagger H = I$:

$$X = \begin{bmatrix} 0 & 1 \\ 1 & 0 \end{bmatrix} \quad X^\dagger = \begin{bmatrix} 0 & 1 \\ 1 & 0 \end{bmatrix} \rightarrow X^\dagger X = XX = I$$

$$H = \frac{1}{\sqrt{2}} \begin{bmatrix} 1 & 1 \\ 1 & -1 \end{bmatrix} \quad H^\dagger = \frac{1}{\sqrt{2}} \begin{bmatrix} 1 & 1 \\ 1 & -1 \end{bmatrix} \rightarrow H^\dagger H = HH = I$$

Unitary Matrices Are Good for Quantum Gates

A question that arises from the previous section: Why go through all the trouble? Why do X and H need to be unitary? The answer is that unitary matrices preserve vector length. This is useful for quantum gates because these require input and output states to be normalized (have a vector length of 1). In fact unitary matrices are the only type of matrices that preserve length and therefore the only type of matrix that can be used for quantum gates. All in all, a deeper question arises: why should quantum gates be linear in the first place and why use a matrix representation at all? We'll try to answer this in a later section, but for now, we'll just have to accept it.

Other Single-Qubit Gates

In the previous section, we saw the single-qubit gates X and H. At the same time, there are other single-qubit gates that are useful in quantum computation.

The X gate has two partners Y, Z. These form the trio known as the Pauli Sigma (σ) gates.

q[0] $|0\rangle$ — X — Y — Z —

$$X = \begin{bmatrix} 0 & 1 \\ 0 & 1 \end{bmatrix}, Y = \begin{bmatrix} 0 & -i \\ i & 1 \end{bmatrix}, Z = \begin{bmatrix} 1 & 0 \\ 0 & -1 \end{bmatrix}$$

These three matrices are useful for information processing tasks such as super dense coding (SDC), a process that seeks to store classical information efficiently in a qubit. They also come up when analyzing atomic properties such as electron spin. Plus they are closely related to the three dimensions of space XYZ.

The rotation gate

q[0] $|0\rangle$ — T —

$$\begin{bmatrix} \cos\theta & -\sin\theta \\ \sin\theta & \cos\theta \end{bmatrix}$$

It is the familiar rotation on real space by an angle θ. This is a unitary matrix, and in this particular case, the T gate performs a $\Pi/4$ rotation around the Z-axis. This gate is required for universal control.

Gates can also manipulate many qubits as we'll see in the next section.

Qubit Entanglement with the Controlled NOT Gate

This gate completes the arsenal of quantum gates required for quantum computation. The controlled NOT (CNOT) is a 2-qubit gate with four computational basis states.

For a superposition, the four basis states CNOT gives

$$\alpha|00\rangle + \beta|01\rangle + \delta|10\rangle + \gamma|11\rangle$$

where α (alpha), β (beta), δ (delta), and γ (gamma) are the superposition coefficients. The quantum circuit is shown as follows:

The matrix representation of CNOT for the basis states is given by

$$
\begin{bmatrix}
1 & 0 & 0 & 0 \\
0 & 1 & 0 & 0 \\
0 & 0 & 0 & 1 \\
0 & 0 & 1 & 0
\end{bmatrix}
\begin{matrix}
|00\rangle \\
|01\rangle \\
|10\rangle \\
|11\rangle
\end{matrix}
$$

The plus (+) symbol is called the target qubit, and the blue dot (below it) is the control qubit. What it does is simple:

- If the control qubit is set to 1, then it flips the target qubit.

- Otherwise it does nothing.

To be more precise, if the first bit is the control, then

$|00\rangle \rightarrow |00\rangle$ *contol* 0 *do nothing*
$|01\rangle \rightarrow |01\rangle$ *contol* 0 *do nothing*
$|10\rangle \rightarrow |11\rangle$ *control* 1 *flip 2nd*
$|11\rangle \rightarrow |10\rangle$ *control* 1 *flip 2nd*

An easy representation of the preceding states is

$|xy\rangle \rightarrow |x\,y \oplus x\rangle$

Tip   The CNOT gate is required to generate entanglement, and it is critical in all kinds of tasks including quantum teleportation, super dense coding, and almost any quantum algorithm out there.

For example, to entangle 2 qubits, apply the Hadamard gate (H) to the first qubit and then apply the CNOT to the second qubit as shown in the following:

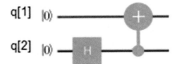

For the basis state in qubit (2), the Hadamard gives

$$|00\rangle \rightarrow \frac{|00\rangle + |10\rangle}{\sqrt{2}}$$

After applying the CNOT, we flip the second qubit if the control is 1, thus

$$|00\rangle \rightarrow \frac{|00\rangle + |11\rangle}{\sqrt{2}}$$

This effectively creates an entangled state between qubits 1 and 2.

All in all, CNOT and single-qubit gates are a powerful arsenal for quantum computation. Because they build up unitary operations on any number of qubits, they are said to be universal for quantum computation. This means that to build a quantum computer that can solve any quantum task, it is enough to use single-qubit gates along with CNOT and measurement gates.

Universal Quantum Computation Delivers Shortcuts over Classical Computation

You may wonder how all the circuits and algebra in the preceding section can help in solving computation tasks that can be easily performed, and probably cheaper, in a classical system. If you consider the so-called bit strength of a classical system

$$x \rightarrow f(x)$$

where, given some input x, the goal is to compute a function f(x) with at least 2^{k-1} elementary operations (where k is the bit strength), then the universal quantum computation can provide an equivalent circuit of roughly the same size that contains the same classical model:

$$|x\rangle, 0 \rightarrow |x\rangle, f(x)$$

What is exciting about the preceding circuit is that there are sometimes shortcuts provided by the greater power of quantum computation to get results faster. This means that you can compute f(x) in fewer than 2^{k-1} operations. For some quantum algorithms such as factorization, the speedups are exponential! This is the true power of quantum systems. So now that you have explored the basic mathematical model of a quantum circuit, it is time to switch into programmatic mode and see how all this can be turned into an actual computer program to be executed on a real quantum device.

Your First Quantum Program

Let's look at the anatomy of a quantum program with a bare-bones example. In this example, we create a single qubit, one classic register to measure the qubit, and then we apply the Pauli X gate (bit flip) on the qubit and finally measure its value. The basic pseudocode of the program can be resumed as follows:

1. Create a quantum program.

2. Create one or more qubits and classical registers to measure the qubits.

3. Create a circuit which groups the qubits in a logical execution unit.

4. Apply quantum gates on the qubits to achieve a desired result.

5. Measure the qubits into the classical register to collect a final result.

6. Compile the program. This step creates a JSON representation of the program in a specific format that will be described later on in this section.

7. Run in the simulator or real quantum device.

8. Fetch the results.

Now let's look at the Python code as well as the Composer circuit in detail.

Listing 4-3. Anatomy of a Quantum Program

```
##############################
import sys
import qiskit
import logging
from qiskit import QuantumProgram

# Main sub
def main():

  # create a  program
  qp = QuantumProgram()

  # create 1 qubit
  quantum_r = qp.create_quantum_register("qr", 1)

  # create 1 classical register
  classical_r = qp.create_classical_register("cr", 1)

  # create a circuit
  qp.create_circuit("Circuit", [quantum_r], [classical_r])

  # get the circuit by name
  circuit = qp.get_circuit('Circuit')

  # enable logging
  qp.enable_logs(logging.DEBUG);

  # Pauli X gate to qubit 1 in the Quantum Register "qr"
  circuit.x(quantum_r[0])

  # measure gate from qubit 0 to classical bit 0
  circuit.measure(quantum_r[0], classical_r[0])

  # backend simulator
  backend = 'local_qasm_simulator'

  # Group of circuits to execute
  circuits = ['Circuit']
```

```
# Compile your program
qobj = qp.compile(circuits, backend)

# run in simulator
result = qp.run(qobj, timeout=240)

# Show result counts
print (str(result.get_counts('Circuit')))

###########################################
# Linux :main()
# windows
if __name__ == '__main__':
  main()
```

Let's see what is going on in Listing 4-3:

- Lines 2–5 import the required libraries: sys (system), qiskit (quantum classes), logging (for debugging), and QuantumProgram: the foundation class for all programs.

- Next, line 11 creates a QuantumProgram. This is the access point to all operations.

- To create a qubit list, use the quantum program create_quantum_register(NAME, SIZE) system call where NAME is the name of the register list and SIZE is the number of qubits. In this case 1 (line 14).

- For each qubit, create a classical register to perform a measurement using the system call create_classical_register(NAME, SIZE).

- Next, create a circuit with the system call create_circuit(NAME, QUANTUM_SET,CLASSIC_SET) where NAME is the name of the circuit, QUANTUM_SET is a list of qubits, and CLASSIC_SET is the list of classic registers. A circuit is the logical unit that holds all qubits and classical registers (line 20).

- Optionally, enable debugging with the system call enable_logs(LEVEL) where LEVEL can be one of logging.DEBUG, logging. INFO, etc. (just the usual logging stuff).

- Next, run the qubit(s) through quantum gates and perform measurements on the qubit(s) to collect results. In this case we apply the Pauli X gate which flips the qubit from its ground state |0> to |1> (lines 25–29).

- Finally, compile the program and run the simulator or real device. In this case we run in the local Python simulator (local_qasm_ simulator) (lines 37–41).

Windows developers watch out! You must wrap your program in a main function and then call it with

```
if __name__ == '__main__':
    main()
```

This is required in Windows because QISKit executes the program using asynchronous tasks (executors), and when the task fires, the subprocess will execute the main module at startup. Thus you need to protect the main code to avoid creating subprocesses recursively. I found this out the hard way when my programs run properly in CentOS but failed in Windows with

```
RuntimeError:
        An attempt has been made to start a new process before the
        current process has finished its bootstrapping phase.

        This probably means that you are not using fork to start your
        child processes and you have forgotten to use the proper idiom
        in the main module:

            if __name__ == '__main__':
                freeze_support()
                ...

        The "freeze_support()" line can be omitted if the program
        is not going to be frozen to produce an executable.
```

This can be a source of grief for the newcomer to Python. Now, run the program to see the output:

```
INFO:qiskit._jobprocessor:<qiskit._result.Result object at
0x000000000D99F470>
{'1': 1024}
```

The result is the JSON document {'1': 1024} where 1 is the measurement of the qubit (remember that we used an X gate to flip the bit) and 1024 is the number of iterations of that result. The probability of this result is calculated by dividing the number of the result iterations (1024) by the total number of iterations of the program (1024). In this case P = 1024/1024 = 1.

Tip   Quantum computers are probabilistic machines. Thus all measurements come attached with a probability for that specific result.

Listing 4-3 can also be described with an equivalent quantum circuit quickly constructed and executed in the IBM Q Experience Composer as shown in Figure 4-4.

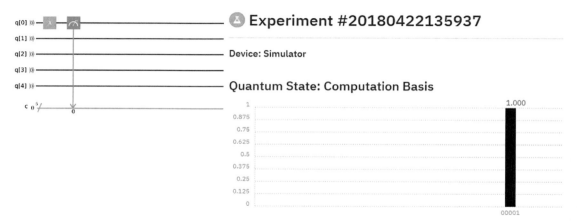

Figure 4-4.   *Composer experiment for Listing 4-3*

Figure 4-4 shows the quantum circuit for Listing 4-3 including the result of the experiment as well as the attached probability. The circuit is very simple as you can see: in the Composer, drag an X gate over qubit 0, then perform a measurement on the same qubit. You will find the Composer a wonderful tool to construct relatively simple circuits, execute them, and visualize their results! Now let's peek into the SDK internals to see how this code gets massaged behind the scenes.

SDK Internals: Circuit Compilation and QASM

Figure 4-5 shows what goes on behind the scenes when your program is run:

- QISKit compiles your program's circuit(s) into a JSON document to be submitted to the local simulator.

- The simulator parses the document, runs the circuit, and returns an opaque JSON document (hidden from the developer).

- QISKit wraps the results JSON document in an object available to the main program. For example, a call to `result.get_counts('Circuit')` extracts the count information from this document.

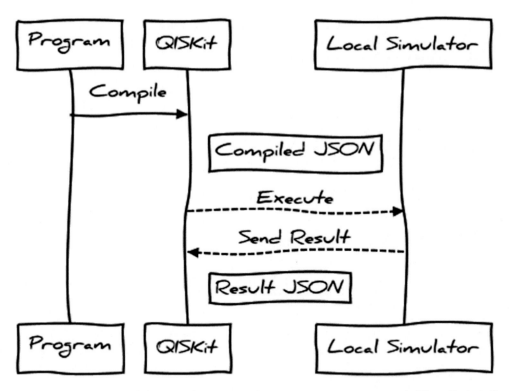

Figure 4-5. *Sequence diagram between the program, QISKit, and local simulator*

Circuit Compilation

Listing 4-4 shows the format of the compiled program before submission to the simulator. The document is made up of

- An execution id

- A header with information about the simulator including name, number of credits used in the execution, plus number of run interactions (shots)

- The circuit section contains an array of circuit objects. Each circuit is made of

  - A circuit name

  - A header (config) with information such as qubit coupling map, basis (physical) gates, runtime seed, and more

  - A compiled circuit section with a header containing information about the qubits and classical registers, as well as an array of operation (or gates) applied to the circuit and their parameters

Listing 4-4. Compilation Format for Listing 4-3

```
{
  "id": "aA46vJHgnKQko3u5L1QqbUDk31sY2m",
  "config": {
    "max_credits": 10,
    "backend": "local_qasm_simulator",
    "shots": 1024
  },
  "circuits": [{
    "name": "Circuit",
    "config": {
      "coupling_map": "None",
      "layout": "None",
      "basis_gates": "u1,u2,u3,cx,id",
      "seed": "None"
    },
```

```
    "compiled_circuit": {
      "operations": [{
        "name": "u3",
        "params": [3.141592653589793, 0.0, 3.141592653589793],
        "texparams": ["\\pi", "0", "\\pi"],
        "qubits": [0]
      }, {
        "name": "measure",
        "qubits": [0],
        "clbits": [0]
      }],
      "header": {
        "number_of_qubits": 1,
        "qubit_labels": [
          ["qr", 0]
        ],
        "number_of_clbits": 1,
        "clbit_labels": [
          ["cr", 1]
        ]
      }
    },
    "compiled_circuit_qasm": "OPENQASM 2.0;\ninclude \"qelib1.inc \";\nqreg
    qr[1];\ncreg cr[1];\nu3(3.14159265358979,0,3.14159265358979) qr[0];\
    nmeasure qr[0] -> cr[0];\n"
  }]
}
```

To display the compiled circuit within your program, print the result of the compilation step as shown in the following command:

```
qobj = qp.compile(circuits, backend)
print(str(qobj))
```

Note   The compilation format is opaque to the programmer and not meant to be accessed directly, but via the SDK API. The reason is that its format may change from version to version. However it is always good to understand what occurs behind the scenes.

Execution Results

This is the response document from the local simulator to the QISKit. The format of this document is shown in Listing 4-5. Remarkable information includes

- Status of the run, execution time, simulator name, and more.

- Result data. This is the information available within your program with the call: `print (str(result.get_counts('Circuit')))`.

Listing 4-5. Results Document from Local Simulator

```
{
  "backend": "local_qiskit_simulator",
  "id": "aA46vJHgnKQko3u5L1QqbUDk31sY2m",
  "result": [{
    "data": {
      "counts": {
        "1": 1024
      },
      "time_taken": 0.0780002
    },
    "name": "Circuit",
    "seed": 123,
    "shots": 1024,
    "status": "DONE",
    "success": true,
    "threads_shot": 4
  }],
```

```
  "simulator": "qubit",
  "status": "COMPLETED",
  "success": true,
  "time_taken": 0.0780002
}
```

Obtaining the results document is a bit trickier because it is an opaque object not exposed to the user's program. Nevertheless you could save the compiled circuit from the previous section and feed it to the simulator manually to obtain the result shown in Listing 4-5. This task is left to you however. The important thing to remember is that the results document (as well as the compilation format) is opaque to the programmer. The reason is that their formats may change over time; nonetheless it is always helpful to understand how things work behind the scenes.

Tip   The compilation and results formats are useful for simulator developers. For example, you could save compilation and results formats for a sample circuit, fix a bug in the C++ simulator, feed it, and compare its results. In this way, your simulator could be easily integrated with the SDK for the rest of us to play with.

Assembly Code

The compiled circuit in Listing 4-4 includes a section that contains a translation of the program into quantum assembly (QASM) as shown in the next paragraph.

```
OPENQASM 2.0;
include "qelib1.inc";
qreg qr[1];
creg cr[1];
x qr[0];
measure qr[0] -> cr[0];
```

Tip   QASM is useful only if running in the remote simulator provided by IBM Q Experience.

QISKit Local Simulators

Access to real quantum devices in IBM Q Experience is restricted by a credit system that diminishes with use; thus we shouldn't run trivial programs such as Listing 4-3. For this purpose QISKit packs a plethora of simulators to satisfy all your testing needs. Table 4-1 provides a list of local and remote simulators available via QISKit and IBM Q Experience by the time of this writing.

Table 4-1. *List of Local and Remote Simulators for IBM Q Experience*

Name	Description
local_qasm_simulator	This is the default Python simulator bundled with QISKit. It is very slow but does the job.
local_clifford_simulator also known as local_qiskit_simulator	A high-performance simulator written in C++ with realistic noise and error simulation.
ibmqx_qasm_simulator	A 24-qubit high-performance remote QASM simulator provided by Q Experience. This is the default remote simulator.
ibmqx_hpc_qasm_simulator	A 32-qubit mega powerful parallel simulator provided by Q Experience. This is a backup to the default remote simulator.

As a simple exercise, obtain a list of IBM Q Experience simulators and real devices by pasting the following REST API URL into your browser:

`https://quantumexperience.ng.bluemix.net/api/Backends?access_token=ACCESS_TOKEN`.

Of course you need an access token which can be easily obtained using the Remote Access API from Chapter 3. Next, let's run our program in other local simulators including the IBM Q Experience remote simulator. Finally let's time our runs to see which simulator is the fastest.

Running in the Local C++ simulator

QISKit uses the pure Python simulator as default (local_qasm_simulator). However you can also use a fast C++ simulator with realistic noise and error rates by changing the backend name of your program to local_clifford_simulator or local_qiskit_simulator (line 35 in Listing 4-3). However there are some caveats you should keep in mind before using it:

- *Linux users*: This simulator uses the C++11 standard which requires gcc 5.3 or later. As a matter of fact the simulator was not built in my CentOS 6 and 7 systems. (you may want to use Windows in this instance).

- *Windows users*: Python uses the CMake utility to build the simulator on the fly. All in all, the default source does not provide a Visual Studio solution to build in Windows. Nevertheless I have taken the time to provide one and fix a couple of crashes I encountered in Windows 7.

Tip   A Windows 64-bit binary for the C++ simulator can be found in the book source under Ch04\qiskit-simulator\qiskit-simulator\x64\Debug. A Visual Studio 2017 solution is also provided if you wish to build it yourself. Make sure you copy all the files in this folder to PYTHON-HOME\Lib\site-packages\qiskit\ backends if missing.

Running in a Remote Simulator

To run in the remote simulator provided by IBM Q Experience, Listing 4-3 needs to change a little bit. Let's see how:

The first thing we need is an IBM Q Experience configuration descriptor with the execution parameters as shown in the next paragraph:

```
APItoken = 'YOU-API-TOKEN'
config = {
    'url': 'https://quantumexperience.ng.bluemix.net/api',
    # The following should only be needed for IBM Q users.
    'hub': 'MY_HUB',
```

```
    'group': 'MY_GROUP',
    'project': 'MY_PROJECT'
}
```

Tip   The preceding code should be kept in a separate file (Qconfig.py) in the same folder as the main program. Get your API token from the IBM Q Experience web console (as shown in Chapter 3) and paste it in the code. Note that hub, group, and project are required for corporate customers only.

Next import the preceding descriptor into the main program:

```
# Q Experience config
import Qconfig

# Main sub
def main():
```

Finally switch the execution backend to the remote simulator:

1. Change the backend name to ibmq_qasm_simulator.

2. Tell the quantum program to use IBM Q Experience by setting the API parameters with the system call: qp.set_api(Qconfig. APItoken, Qconfig.config['url']) where APItoken and URL are the values from the configuration descriptor.

3. Execute in IBM Q Experience with the system call: result = qp.execute(circuits, backend, shots=512, max_ credits=3). Note that we do not compile and run the circuit as before. Therefore you must remove the calls to qobj = qp.compile(circuits, backend) and result = qp.run(qobj, wait=2, timeout=240).

The changes are shown in the following command. Make sure you remove the old compilation and run calls or the program will fail:

```
backend = 'ibmqx_qasm_simulator'

# Group of circuits to execute
circuits = ['Circuit']
```

```
# set the APIToken and Q Experience API url
qp.set_api(Qconfig.APItoken, Qconfig.config['url'])

result = qp.execute(circuits, backend, shots=512, max_credits=3, wait=10,
timeout=240)
```

Finally, execute and test. The output should look something like

```
DEBUG:qiskit.backends._qeremote:Running on remote backend ibmq_qasm_
simulator with job id: 3677ff592e5e5a6fd31a569b0b4faf92
INFO:qiskit._jobprocessor:<qiskit._result.Result object at
0x0000000004A35160>
{'1': 512}
```

Now let's put all this together and see who the fastest simulator is. My money is in C++.

And the Fastest Simulator Is Comparing Execution Times

I have gathered execution times for all simulators in an x64 machine running Windows 7. Incredibly, the fastest simulator turned out to be the IBM Q Experience remote, followed closely by the pure Python, and lastly my personal favorite: C++ (see Figure 4-6).

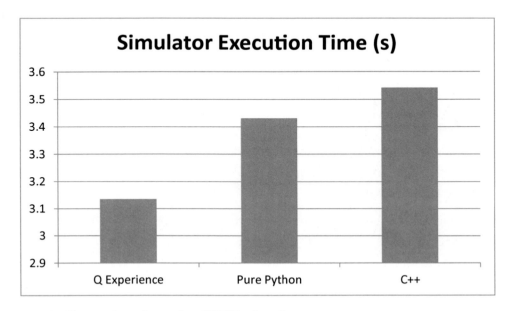

Figure 4-6. *Execution times for QISKit simulators*

Even though the call goes through the network, the IBM Q Experience remote simulator manages to outperform the others. What I found perplexing is how an interpreted Python simulator can be faster than a native code implementation. This is probably due to the fact that the native invocation uses asynchronous tasks to spawn the C++ simulator process, thus slowing things down enough for the Python code to outperform it. Now that you have learned how to run a program in the simulator, let's do it in the real thing.

Running in a Real Quantum Device

Let's modify the program from the previous section to make a more complex circuit instead. Listing 4-6 shows a sample circuit that performs a series of rotations on the first qubit of a quantum computer. The rotations demonstrate the use of the physical gates of the real quantum processor ibmqx4: u1, u2, and u3 to rotate a single qubit over the X-, Y-, and Z-axis of the Bloch sphere by theta, phi, or lambda degrees.

Tip   The Bloch sphere is the geometrical representation of a single qubit where the top of the Z-axis represents the basis state I0> and the bottom I1>. A rotation over a given axis represents the probability that the qubit will collapse in certain direction when a measurement is performed (see Figure 4-7).

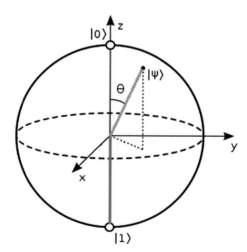

Figure 4-7. *Bloch sphere representation of a qubit*

175

Physical gates (also known as basis gates) are important because they constitute the foundation under which more complex logical gates are constructed. Hence Listing 4-6 performs the following steps:

- It allocates 5 qubits and five classical measurement registers corresponding to the 5 qubits available from the ibmqx4 processor in Q Experience (lines 17–20).

- Next, a sequence of rotations on the first qubit are performed using the basis gates u1, u2, and u3 (lines 29–34).

- Finally, a measurement is performed in the qubit and the result stored in the classical register.

- Before execution the backend is set to ibmqx4 (a 5-qubit processor – line 42), and the authentication token and API URL are set via set_api(Qconfig.APItoken, Qconfig.config['url']).

- To execute in the real quantum device, use the QuantumProgram execute system call execute(NAMES, BACKEND, shots=SHOTS, max_credits=CREDITS, timeout=TIMEOUT) where

  - NAMES is a list of circuit names.

  - SHOTS is the number of iterations performed in the circuit. The higher the number, the greater the accuracy.

  - CREDITS is the maximum number of points that you wish to be deducted from your execution bank (15 is the default startup number). Note that the more shots are performed, the more credits will be deducted from your bank. Keep this in mind before you run out of credits.

  - TIMEOUT is the read timeout from the remote end point.

Note   Python quantum programs/experiments executed against a real device are not recorded in the Composer-Scores section of IBM Q Experience. This is because Python uses the Jobs REST API behind the scenes which puts the experiment in an execution queue instead. If you wish to record your executions in the Composer, you could use the web console or REST APIs as shown in the next section.

Listing 4-6. Sample Circuit #2

```python
import sys,time,math
import qiskit
import logging
from qiskit import QuantumProgram

# Q Experience config
import Qconfig

# Main sub
def main():

  # create a  program
  qp = QuantumProgram()

  # create 1 qubit
  quantum_r = qp.create_quantum_register("qr", 5)

  # create 1 classical register
  classical_r = qp.create_classical_register("cr", 5)

  # create a circuit
  circuit = qp.create_circuit("Circuit", [quantum_r], [classical_r])

  # enable logging
  qp.enable_logs(logging.DEBUG);

  # first physical gate: u1(lambda) to qubit 0
  circuit.u2(-4 *math.pi/3, 2 * math.pi, quantum_r[0])
  circuit.u2(-3 *math.pi/2, 2 * math.pi, quantum_r[0])
  circuit.u3(-math.pi, 0, -math.pi, quantum_r[0])
  circuit.u3(-math.pi, 0, -math.pi/2, quantum_r[0])
  circuit.u2(math.pi, -math.pi/2, quantum_r[0])
  circuit.u3(-math.pi, 0, -math.pi/2, quantum_r[0])

  # measure gate from qubit 0 to classical bit 0
  circuit.measure(quantum_r[0], classical_r[0])
  circuit.measure(quantum_r[1], classical_r[1])
  circuit.measure(quantum_r[2], classical_r[2])

  # backend
  backend = 'ibmqx4'
```

177

```python
# Group of circuits to execute
circuits = ['Circuit']

# set the APIToken and Q Experience API url
qp.set_api(Qconfig.APItoken, Qconfig.config['url'])

result = qp.execute(circuits, backend, shots=512, max_credits=3,
timeout=240)

# Show result counts
print ("Job id=" + str(result.get_job_id()) + " Status:" + result.get_status())

#############################################
if __name__ == '__main__':
  start_time = time.time()
  main()
  print("--- %s seconds ---" % (time.time() - start_time))
```

Quantum Circuit for the Composer

The program in Listing 4-6 can also be created in the IBM Q Experience Composer using their slick drag and drop user interface. Simply drag the gates into the qubit histogram as shown in Figure 4-8, set the parameters for the gate(s), and finally save and run in the simulator or real device.

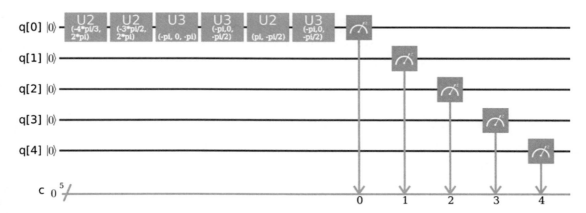

Figure 4-8. *Q Experience Composer circuit for Listing 4-6*

For those of you who prefer the raw power of assembly, the Composer allows to copy-paste code directly into the console in assembly mode as shown in Figure 4-9. It will even parse any syntax errors in your code and show you the offending line(s).

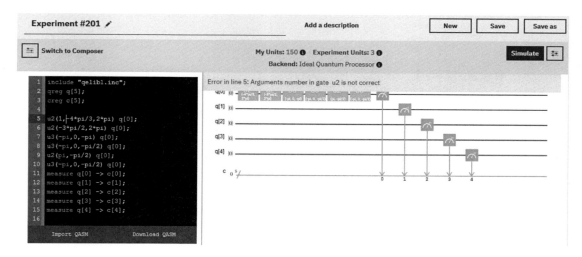

Figure 4-9. *Composer in assembly mode for circuit in Figure 4-8*

There are multiple ways of executing your experiment in IBM Q Experience; one of the most interesting is using their awesome REST API.

Execution via Your Favorite REST Client

This is one of the most exciting ways to interact with Q Experience. By using simple REST requests, you can do pretty much anything you do in Python or the Composer:

- List backend devices.

- List hardware or calibration parameters for the real devices.

- Get information about the job execution queue.

- Get the status of a job or experiment.

- Push or cancel jobs.

- Execute an experiment and record it in the Scores section of the Composer.

Tip   The REST API allows you to use any language to create your own interface to Q Experience (even a web browser). This API is described in full detail in Chapter 3.

There are two ways of submitting experiments using REST: via the jobs and the execute APIs. Let's see how.

Run via the Jobs API

You can use your favorite browser REST client to submit the experiment in Listing 4-6. For example, using Chrome's YARC (Yet Another REST Client) create an HTTP POST request to the end point:

```
https://quantumexperience.ng.bluemix.net/api/Jobs?access_token=ACCESS_TOKEN
```

The tricky part is getting your access token or access key. For this part you must authenticate using your API token or username and password. Note that the API token is not to be confused with the access token. To obtain an access token, you must do an authentication request. (Take a look at Chapter 3 under "Remote Access via the REST API.")

Tip   Chrome's YARC allows you to construct REST requests and save them as favorites. Create an authentication request to IBM Q Experience as described in Chapter 3, save it as a favorite, and use it every time to obtain an access token to test other REST API calls.

The request payload is a JSON document shown in Listing 4-7. The format is described in Table 4-2.

Table 4-2. *Request Format for the Jobs API*

Key	Description
qasms	This is an array of assembly code programs all in one line separated by the line feed character (\n).
Shots	The number of iterations you code will go through.
backend	This is an object that describes the backend. In this case ibmqx4.
maxCredits	This is a hint of the number of credits to be deducted from your account balance.

Listing 4-7. HTTP Request for the Jobs API

```
{
  "qasms": [{
    "qasm": "\n\ninclude \"qelib1.inc\";\nqreg q[5];\ncreg c[5];\nu2
    (-4*pi/3,2*pi) q[0];\nu2(-3*pi/2,2*pi) q[0];\nu3(-pi,0,-pi) q[0];\nu3
    (-pi,0,-pi/2) q[0];\nu2(pi,-pi/2) q[0];\nu3(-pi,0,-pi/2) q[0];\nmeasure
    q -> c;\n"
  }],
  "shots": 1024,
  "backend": {
    "name": "ibmqx4"
  },
  "maxCredits": 3
}
```

Once you have obtained an access token, copy-paste the payload from Listing 4-7 into your REST client, submit, and wait for a response. If all goes well, you should see a response similar to Listing 4-8.

Listing 4-8. HTTP Response from Q Experience

```json
{
  "qasms": [
    {
      "qasm": "\n\ninclude \"qelib1.inc\";\nqreg q[5];\ncreg c[5];\nu2
      (-4*pi/3,2*pi) q[0];\nu2(-3*pi/2,2*pi) q[0];\nu3(-pi,0,-pi) q[0];\
      nu3(-pi,0,-pi/2) q[0];\nu2(pi,-pi/2) q[0];\nu3(-pi,0,-pi/2) q[0];\
      nmeasure q -> c;\n",
      "status": "WORKING_IN_PROGRESS",
      "executionId": "e9d758c3480a54a6455f72c84c5cc2a6"
    }
  ],
  "shots": 1024,
  "backend": {
    "id": "c16c5ddebbf8922a7e2a0f5a89cac478",
    "name": "ibmqx4"
  },
  "status": "RUNNING",
  "maxCredits": 3,
  "usedCredits": 3,
  "creationDate": "2018-04-24T00:12:07.847Z",
  "deleted": false,
  "id": "33d58594fcb7204e4d2ccdb65cd3c88c",
  "userId": "ef072577bd26831c59ddb212467821db"
}
```

A partial response format is described in Table 4-3.

Table 4-3. *Response Format for the Jobs API*

Key	Description
qasms	An array of objects that includes The submitted code. The runtime status: WORKING_IN_PROGRESS, COMPLETED, or FAILED. An execution id for the code.
shots	The number of iterations of the experiment.
backend	An object with information about the backend such as name and id.
status	The overall status of the job: RUNNING, COMPLETED, or FAILED.
maxCredits	The maximum number of credits used for this run.
usedCredits	The actual number of credits spent in this run.
creationDate	Date the job was created.
deleted	True if a request to delete the job has been submitted, else false. Note: Canceled or deleted jobs will linger for a while before being purged from the queue.
id	Id of this job.
userId	User id of the owner.

Tip   The jobs (as well as the execute) APIs are undocumented and not meant to be accessed directly at this point. Thus the response format may vary over time. Perhaps this will change in the future and the REST API will be part of the official SDK. In the meantime however your results may be different from mine.

Run via the Execute API

The main difference between this and the jobs APIs is that the execute API registers the experiment in the Composer. To see how this is done, create an HTTP POST request to the end point:

```
https://quantumexperience.ng.bluemix.net/api/codes/execute?access_token=TOK
EN&shots=1&seed=SEED&deviceRunType=ibmqx4
```

The arguments to the request are

- *access_token*: Your access token

- *shots*: The number of iterations of the experiment

- *seed*: A random execution seed required only if running in the simulator

- *deviceRunType*: The name of the device where the experiment will be run

The request payload is shown in Listing 4-9. Every experiment must include a name. The code type is QASM2, and the assembly code must be written in a single line separated by a line feed (\n).

Listing 4-9. HTTP Request Payload for the Execution API

```
{
  "name": "Experiment #20180410193125",
  "codeType": "QASM2",
  "qasm": "\n\ninclude \"qelib1.inc\";\nqreg q[5];\ncreg c[5];\nu2
  (-4*pi/3,2*pi) q[0];\nu2(-3*pi/2,2*pi) q[0];\nu3(-pi,0,-pi) q[0];\nu3
  (-pi,0,-pi/2) q[0];\nu2(pi,-pi/2) q[0];\nu3(-pi,0,-pi/2) q[0];\nmeasure
  q -> c;\n"
}
```

Submit the request using your REST client and wait for a result. Listing 4-10 shows a reduced response format for the experiment.

Tip   Save yourself a lot of headaches. Always make sure the device is online and the qasm is written in a single line including line feeds (\n) before submission or you will have a lot of trouble. Double- and triple-check this or your request will fail most of the time.

Listing 4-10. Response Format for the Execute API

```
{
  "startDate": "2018-04-24T22:31:23.555Z",
  "modificationDate": 1524609083555,
  "typeCredits": "plan",
  "status": {
    "id": "WORKING_IN_PROGRESS"
  },
  "deviceRunType": "real",
  "ip": {
    "ip": "172.58.152.206",
    "country": "United States",
    "continent": "North America"
  },
  "shots": 1,
  "paramsCustomize": {},
  "deleted": false,
  "userDeleted": false,
  "id": "1203b1158e6ae537e8b770cb8049a6ae",
  "codeId": "e0f5c573eef75581cf16bce4187ecab8",
  "userId": "ef072577bd26831c59ddb212467821db",
  "infoQueue": {
    "status": "PENDING_IN_QUEUE",
    "position": 108
  },
  "code": {
    "type": "Algorithm",
    "active": true,
    "versionId": 1,
    "idCode": "e86d38c389f4449e62756922a1aa5729",
    "name": "Experiment #201",
    "jsonQASM": {
      "gateDefinitions": [],
      "topology": "3b8e671a5a3b56899e6e601e6a3816a1",
      "playground": [
```

```json
    {
      "name": "q",
      "line": 0,
      "gates": [
            ...
      ]
    },
    {
      "name": "q",
      "line": 4,
      "gates": [
        {
          "name": "measure",
          "qasm": "measure",
          "position": 10,
          "measureCreg": {
            "line": 5,
            "bit": 4
          }
        }
      ]
    },
    {
      "name": "c",
      "line": 0
    }
  ],
  "numberGates": 7,
  "hasMeasures": true,
  "numberColumns": 11,
  "include": "include \"qelib1.inc\";"
},
"qasm": "\n\ninclude \"qelib1.inc\";\nqreg q[5];\ncreg c[5];\nu2
(-4*pi/3,2*pi) q[0];\nu2(-3*pi/2,2*pi) q[0];\nu3(-pi,0,-pi) q[0];\nu3
(-pi,0,-pi/2) q[0];\nu2(pi,-pi/2) q[0];\nu3(-pi,0,-pi/2) q[0];\nmeasure
q -> c;\n",
```

```
      "codeType": "QASM2",
      "creationDate": "2018-04-24T22:31:22.561Z",
      "deleted": false,
      "orderDate": 1524609083391,
      "userDeleted": false,
      "isPublic": false,
      "id": "e0f5c573eef75581cf16bce4187ecab8",
      "userId": "ef072577bd26831c59ddb212467821db"
   }
}
```

There is a lot of information returned by this response, and most of the data is straightforward. Nevertheless Table 4-4 describes the most important values.

Table 4-4. *Miscellaneous Information Returned by the Execute API*

Key	Description
status	The status of the execution. It can be one of the following: WORKING_IN_PROGRESS, COMPLETED, or FAILED.
deviceRunType	The device where the experiment has been run: real (for real devices) or simulator.
infoQueue	Information about the execution queue including • The status: PENDING_IN_QUEUE. • Position in the queue.
code	A very detailed description of the experiment including • Quantum gates, parameters, position, and more. • Assembly code. • Miscellaneous information such as name, type, status, version, and others.

Tip After receiving a response, log in to the IBM Q Experience console. The experiment name should be displayed in the Quantum Scores section of the Composer.

Quantum Assembly: The Power Behind the Scenes

You have probably realized what goes behind the scenes when an experiment is executed within the Composer or a REST client. The circuit gets translated into quantum assembly (QASM) and then executed in the real device or simulator. Quantum assembly is an intermediate representation of the high-level Python code and is the result of the collaboration between IBM Q Experience and the open source community.

Tip   QASM is based on its classical cousin which has become sort of a lost art. It is not as scary as its cousin though. As a matter of fact, it is really based on a subset of the classical assembly grammar.

Formally, the life cycle of your Python program or Q Experience circuit can be described as a cross between quantum and classical parts of a computation with the following steps:

- *Compilation*: This is an offline step that takes place in a classical computer. When a Python or Composer circuit runs, the classical compiler translates a high-level representation (e.g., Python) into the QASM intermediate representation. This step has the following characteristics:

  - Specific problem parameters are not yet known.

  - No interaction with the quantum computer is required.

  - It is possible to compile classical procedures into object code and make initial optimizations. For example, the Python program in Listing 4-6 and corresponding Composer circuit are translated into the assembly shown in Listing 4-11.

Listing 4-11.   QASM Code for Python in Listing 4-6

```
include "qelib1.inc";
qreg qr[5];
creg cr[5];
u2(-4.18879020478639,6.28318530717959) qr[0];
u2(-4.71238898038469,6.28318530717959) qr[0];
```

```
u3(-3.14159265358979,0,-3.14159265358979) qr[0];
u3(-3.14159265358979,0,-1.57079632679490) qr[0];
u2(3.14159265358979,-1.57079632679490) qr[0];
u3(-3.14159265358979,0,-1.57079632679490) qr[0];
measure qr[0] -> cr[0];
measure qr[1] -> cr[1];
measure qr[2] -> cr[2];
```

- *Circuit generation*: The QASM from the previous step gets fed to the circuit generation phase. This step takes place on a classical computer where the specific problem parameters are known, and some interaction with the quantum computer may occur. This step has the following characteristics:

  - This is an online phase (occurs in a quantum computer).

  - The output is a collection of quantum circuits, or quantum basic blocks, together with associated classical control instructions and classical object code needed at runtime.

- *Execution*: This step takes place on a physical quantum computer. The input is a collection of quantum circuits expressed using a quantum circuit intermediate representation. These are executed on a low-level controller, and the output is a collection of measurement results returned from the high-level controller.

- *Postprocessing*: This step takes place on a classical computer and receives a collection of processed measurement results. The output is the final result of the quantum computation (see Figure 4-10).

189

Device: Simulator

Quantum State: Computation Basis

Quantum Circuit

Figure 4-10. *Postprocessing result from the circuit life cycle for Listing 4-6*

All in all, quantum assembly syntax is not as scary as its classical counterpart; as a matter of fact, programming in quantum assembly directly turns out to be simpler and faster than using Python. The next section presents a set of simple tricks to use if you decide to code directly in QASM:

- Always begin by including the header `include "qelib1.inc"`. It contains Q Experience hardware primitives (quantum gates). The gates provided in this library are described in Table 4-5 for single-qubit gates and Table 4-6 for multiqubit gates.

Table 4-5. *Single-Qubit Gates Provided by Quantum Assembly*

Name	Description
u3(theta,phi,lambda)	3-parameter 2-pulse single qubit.
u2(phi,lambda)	2-parameter 1-pulse single qubit.
u1(lambda)	1-parameter 1-pulse single qubit.
Id	Equivalent to the identity matrix or u(0,0,0).
X	Pauli X or σ_x (sigma-x) or bit flip.
Y	Pauli Y or σ_y (sigma-y).
Z	Pauli Z or σ_z (sigma-y).
rx(theta)	Rotation around X-axis by theta degrees.
ry(theta)	Rotation around Y-axis by theta degrees.
rz(Phi)	Rotation around Z-axis by theta degrees.
H	Hadamard: Puts a single qubit in superposition of states.
S	Square root of Z: sqrt(Z) phase gate.
Sdg	S-dagger: The complex conjugate of S. Algebraically it is defined as the complex conjugate of the transpose matrix of sqrt(z).
T	The sqrt(S) phase gate.
Tdg	T-dagger or the complex conjugate of sqrt(S).

Table 4-6. *Multiqubit Gates Provided by Quantum Assembly*

Name	Description
cx c,t	Controlled NOT (CNOT): It flips the second qubit (t) only if the control qubit (c) is 1. It is used to entangle 2 qubits.
cz a,b	Controlled phase: Applies a phase rotation only if the control qubit (a) is 1.
cy a,b	Controlled Y: Applies a Pauli Y rotation only if the control qubit (a) is 1.
ch a,b	Controlled H: Puts qubit (b) in superposition only if control qubit (a) is 1.
ccx a,b,c	3-qubit Toffoli gate: It flips qubit c only if qubits a and b are 1.

- Declaring qubit registers (arrays) is simple. For example, to declare a register consisting of 5 qubits: `qreg qr[5];` Note: All instructions are separated by semicolon.

- To declare a register consisting of 5 classical bits, use `creg cr[5];`

- To apply a gate to a specific qubit, simply type the gate name and the target qubit. For example, to put the first qubit in superposition (for a quantum number generator), use `h q[0];`

- The final step of your program should always be to perform a measurement on the qubit. For example, to measure our superimposed qubit and store it in the first classical register, use `measure qr[0] -> cr[0];`

Note that quantum computers are probabilistic machines; therefore the definite state of the qubit cannot be known (it is forbidden by quantum mechanics). Thus all we get is the probability that the qubit is in state 0 or 1. For the simple quantum number generator on qubit zero `h q[0]` in the preceding paragraph, we can use the probability of state 1 as our random number. This can be seen as a neat graph shown by the IBM Q Experience Composer when the results are collected after the assembly code executes as shown in Figure 4-9.

You have taken the first step in this new career as a quantum programmer in the cloud. By using the high-level Python SDK and powerful quantum assembly engine, experiments can be run in the awesome IBM Q Experience platform. These skills will be valuable in a few years when quantum computers start to join the data center. In the next chapter, we take things to the next level with a set of algorithms that show the almost magical powers of quantum mechanics when applied to computation. So read on.

Start Your Engines: From Quantum Random Numbers to Teleportation, Pit Stop at Super Dense Coding

This chapter takes you through a journey about three remarkable information processing capabilities of quantum systems. We start with one of the simplest procedures by exploring the fundamentally random nature of quantum mechanics as a source of true randomness. Next, the chapter looks at perhaps two exuberant but related procedures called super dense coding and quantum teleportation. In super dense coding, you will learn how it is possible to send 2 classical bits of information using a single qubit. In quantum teleportation, you will learn how the quantum state of a qubit can be recreated by a hybrid classical-quantum information transfer procedure. All algorithms include circuit design for the IBM Q Experience Composer as well as Python and QASM code. Results will be gathered for display and analysis, so let's get started.

Quantum Random Number Generation

In this section you will learn how the probabilistic nature of a quantum computer can be exploited to generate random bits or numbers using the Hadamard gate.

© Vladimir Silva 2018
V. Silva, *Practical Quantum Computing for Developers*, https://doi.org/10.1007/978-1-4842-4218-6_5

Random Bit Generation Using the Hadamard Gate

Hadamard is one of the fundamental gates in any quantum information system. It is used to put a qubit in a superposition of states. Algebraically, it is described by the matrix

$$H = \frac{1}{\sqrt{2}}\begin{bmatrix} 1 & 1 \\ 1 & -1 \end{bmatrix}$$

To understand better how this matrix puts a qubit in superposition, consider the geometrical representation of a single qubit:

In Figure 5-1 the basis states of the qubit are described using ket notation where $|0\rangle = \begin{bmatrix} 1 \\ 0 \end{bmatrix}$ and $|1\rangle = \begin{bmatrix} 0 \\ 1 \end{bmatrix}$. Remember from the previous chapter that a ket is simply a unitary vector (a vector of length 1). Thus the general (or superposition) state is then defined by the unitary vector $\psi = \alpha|0\rangle + \beta|1\rangle$ where α and β are complex coefficients. Applying H to the basis states gives

$$H|0\rangle = \frac{1}{\sqrt{2}}\begin{bmatrix} 1 & 1 \\ 1 & -1 \end{bmatrix}\begin{bmatrix} 1 \\ 0 \end{bmatrix} = \frac{1}{\sqrt{2}}\begin{bmatrix} 1 \\ 1 \end{bmatrix} = \frac{1}{\sqrt{2}}\left(\begin{bmatrix} 1 \\ 0 \end{bmatrix} + \begin{bmatrix} 0 \\ 1 \end{bmatrix}\right) = \frac{|0\rangle + |1\rangle}{\sqrt{2}}$$

$$H|1\rangle = \frac{1}{\sqrt{2}}\begin{bmatrix} 1 & 1 \\ 1 & -1 \end{bmatrix}\begin{bmatrix} 0 \\ 1 \end{bmatrix} = \frac{1}{\sqrt{2}}\begin{bmatrix} 1 \\ -1 \end{bmatrix} = \frac{1}{\sqrt{2}}\left(\begin{bmatrix} 1 \\ 0 \end{bmatrix} - \begin{bmatrix} 0 \\ 1 \end{bmatrix}\right) = \frac{|0\rangle - |1\rangle}{\sqrt{2}}$$

And for the superimposed state ψ

$$\Psi = \alpha|0\rangle + \beta|1\rangle \rightarrow \alpha\left(\frac{|0\rangle + |1\rangle}{\sqrt{2}}\right) + \beta\left(\frac{|0\rangle - |1\rangle}{\sqrt{2}}\right) = \frac{\alpha + \beta}{\sqrt{2}}|0\rangle + \frac{\alpha - \beta}{\sqrt{2}}|1\rangle$$

All in all, the Hadamard gate expands the range of states that are possible for a quantum circuit. This is important because the expansion of states creates the possibility of finding shortcuts resulting in faster computation.

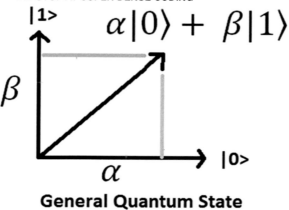

General Quantum State

Figure 5-1. *Geometrical representation of the general (superimposed) state* ψ *of a qubit*

Tip   Quantum mechanics says that we can't predict with certainty the values of coefficients α and β in the preceding basis states, even given complete knowledge of the laws of physics or a particle's initial conditions. The best we can do is to calculate a probability.

With this in mind, a random bit generator circuit implementation is as simple as it gets. In the IBM Q Experience Composer, create a circuit with a Hadamard gate for the first qubit, and then perform a measurement in the basis state as shown in Figure 5-2.

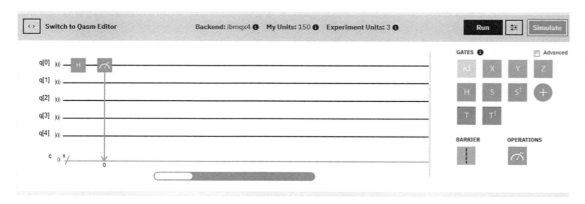

Figure 5-2. *Circuit for a random bit generation*

It is probably not a good idea to run this in the real device as it may take a while (remember that executions are scheduled and may take time depending on the number of jobs in the run queue). Plus each execution in a real device depletes your credits. Run the circuit in the simulator to obtain an immediate result (see Figure 5-3). Note that each outcome (0 or 1) has an equal probability of ½, thus we can create random bits given: probability for outcome 1 > ½ (got 1) else (got 0).

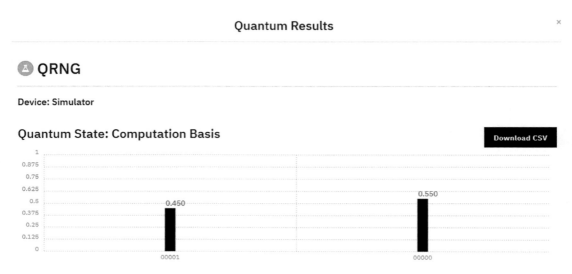

Figure 5-3. *Execution results for circuit in Figure 5-2*

Of course this is a very inefficient way of generating random bits. A better way would be to write a QISKit Python script to programmatically create a circuit to do the job. Listing 5-1 shows a simple script to generate n random numbers using x qubits where the number of bits is 2^x. By default, the script generates 10 8-bit random numbers using 3 qubits, that is, n = 10 and x = 3, given $2^3 = 8$. Let's take a closer look:

- Line 12 defines the function qrng to create a circuit using n qubits.

- Using the QISKitAPI, lines 15-21 create a QuantumProgram with n qubits and n classical registers to store the measurements.

- A Hadamard gate is applied to all qubits, then a measurement is performed on each, and finally the result is stored in classical register n (lines 30-35).

- The circuit is compiled to run in the Q Experience remote simulator by using the system call set_api(API-TOKEN, URL). Note that you will need your configuration descriptor with the API token and end point URL. The circuit gets executed and the result counts are collected (lines 40-51).

- Finally to generate random bits, look at the outcome counts. For example, given the results {'100': 133, '101': 134, '011': 131, '110': 125, '001': 109, '111': 128, '010': 138, '000': 126}. For each outcome, if the count is greater than the average probability, then you get a 1, else you get a 0. The average probability is calculated by dividing the number of shots (1024 in this case) by the number of outcomes (2^x where x is the number of qubits (default is 3) – 1024/8 = 128). Thus, for the preceding results

133	1	
134	1	11100010 = 226
131	1	
125	0	
109	0	
128	0	
138	1	
126	0	

Listing 5-1. Quantum Program to Generate n Random Numbers of 2^x Bits

```
#############################
import sys,time
import qiskit
import logging
from qiskit import QuantumProgram

# Q Experience config
sys.path.append('../Config/')
import Qconfig
```

```python
# Generate an 2**n bit random number where n = # of qubits
def qrng(n):

  # create a  program
  qp = QuantumProgram()

  # create n qubit(s)
  quantum_r = qp.create_quantum_register("qr", n)

  # create n classical registers
  classical_r = qp.create_classical_register("cr", n)

  # create a circuit
  circuit = qp.create_circuit("QRNG", [quantum_r], [classical_r])

  # enable logging
  #qp.enable_logs(logging.DEBUG);

  # Hadamard gate to all qubits
  for i in range(n):
    circuit.h(quantum_r[i])

  # measure qubit n and store in classical n
  for i in range(n):
    circuit.measure(quantum_r[i], classical_r[i])

  # backend simulator
  backend = 'ibmq_qasm_simulator'

  # Group of circuits to execute
  circuits = ['QRNG']

  # Compile your program: ASM print(qp.get_qasm('Circuit')), JSON:
  print(str(qobj))
  # set the APIToken and Q Experience API url
  qp.set_api(Qconfig.APItoken, Qconfig.config['url'])
  shots=1024
  result = qp.execute(circuits, backend, shots=shots, max_credits=3,
  timeout=240)
```

```
# Show result counts
# counts={'100': 133, '101': 134, '011': 131, '110': 125, '001': 109,
'111': 128, '010': 138, '000': 126}
counts = result.get_counts('QRNG')
bits = ""
for v in counts.values():
  if v > shots/(2**n) :
    bits += "1"
  else:
    bits += "0"

return int(bits, 2)


###########################################
if __name__ == '__main__':
  start_time = time.time()
  numbers = []

  # generate 100 8 bit rands
  size = 10
  qubits = 3 # bits = 2**qubits

  for i in range(size):
    n = qrng(qubits)
    numbers.append(n)

  print ("list=" + str(numbers))
  print("--- %s seconds ---" % (time.time() - start_time))
```

Caution   Before executing any program, always make sure your configuration
is correct including a valid API token and end point URL. This is a major source of
headaches. Remember that your program will fail if you miss this crucial step.

A quantum circuit for Listing 5-1 is shown in Figure 5-4. The circuit uses 3 qubits to
generate an 8-bit random number between 0 and 255.

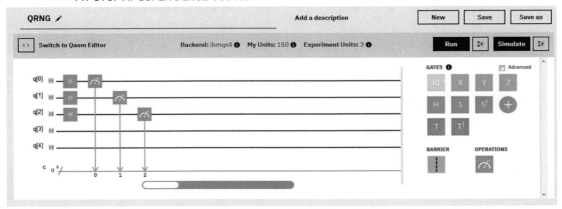

Figure 5-4. *Q Experience circuit for Listing 5-1*

Let's gather some data from multiple runs and put the results to the test.

Putting Randomness Results to the Test

Linux provides a neat program called ent (short for entropy) which is called a pseudorandom number sequence test program.[1] We can use this command to test the numbers generated in the previous section.

Tip   Windows users – a Windows 32 binary is available for download from the project site. A binary is also included in the source for this chapter under Workspace\Ch05\ent.exe.

Thus I have gathered around 200 random 8-bit numbers generated using Listing 5-1. Using ent, this sequence can be put to the test with the command *ent [infile]* as shown in the next paragraph.

```
C:\Workspace\Ch05>ent qrnd-stdout.txt
Entropy = 3.122803 bits per byte.
Optimum compression would reduce the size of this 805 byte file by 60 percent.

Chi square distribution for 805 samples is 29149.54, and randomly would
exceed this value less than 99.9 percent of the times.
```

---

[1]ENT – A Pseudorandom Number Sequence Test Program available at http://fourmilab.ch/random/

Arithmetic mean value of data bytes is 46.1503 (127.5 = random).
Monte Carlo value for Pi is 4.000000000 (error 27.32 percent).
Serial correlation coefficient is -0.356331 (totally uncorrelated = 0.0).

According to the authors, the Chi-square test determines the quality of the random sequence. If the Chi-square percent distribution is less than 1% or greater than 99%, then the sequence is no good. My output shows a percentage of 99.9% which indicates the randomness of the numbers is low. This is probably due to the fact that I used the remote simulator. This simulator is probably based on the default UNIX random number generator (a poor-quality generator). See if your sequence does any better. The next table shows the results from various deterministic and quantum sources head to head provided by the developers of ENT.[2]

Table 5-1. *Randomness Test Results from Various Sources Gathered by ENT[1]*

Source	Chi-square percentage
UNIX rand()	99.9% for 500,000 samples (bad)
Improved UNIX generator by Park & Miller	97.53% for 500,000 samples (better)
HotBits: random numbers, generated by radioactive decay	40.98% for 500,000 samples (the best)

The preceding table clearly shows that UNIX rand() shouldn't be trusted for random number generation. If you need lots of truly random numbers (e.g., to generate encryption keys), use a quantum source such as HotBits. All in all, the purpose of this section has been to get your feet wet with a simple quantum circuit for random number generation. The next section takes things to the next level with the bizarre quantum data transfer protocol dubbed super dense coding.

Super Dense Coding

Super dense coding (SDC) is a data transfer protocol that demonstrates the remarkable information processing capabilities of a quantum system. Formally, SDC is a simple procedure that allows for transferring 2 classical bits of information to another party using a single qubit. The protocol is illustrated in Figure 5-5.

---

[2]HotBits: Genuine random numbers, generated by radioactive decay available online at http://fourmilab.ch/hotbits/.

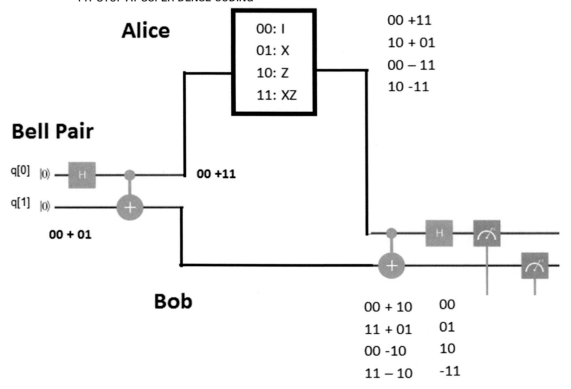

Figure 5-5. Super dense coding protocol

1. The process starts with a third party (Eve) generating what is
   called a Bell Pair. Eve starts with 2 qubits in the basis state |0>. She
   applies a Hadamard gate to the first qubit to create superposition.
   It then applies a CNOT gate using the first qubit as the control
   (dot) and the second as the target (+). This results in the states
   shown in Table 5-2.

Table 5-2. *Bell Pair States*

Gate	Outcome states	Details
H	$\lvert 00\rangle \rightarrow \lvert 00\rangle + \lvert 10\rangle$	When the H gate is applied to the first qubit, it enters superposition; thus we get the states 00 + 10 where the second qubit remains as 0. Note that the square root (2) from the Hadamard matrix has been omitted for simplicity.
CNOT	$\lvert 00\rangle + \lvert 10\rangle \rightarrow \lvert 00\rangle + \lvert 11\rangle$	The CNOT gate entangles both qubits. In particular, it flips the target (+) if the control (.) is 1, else it leaves intact. Thus we flip the second qubit if the first is 1 resulting in 00 + 11.

2. In the second step of the process, the first qubit is sent to Alice and the second to Bob. Note that Alice and Bob may be in remote places. The goal of the protocol is for Alice to send 2 classical bits of information to Bob using her qubit. But before she does, she needs to apply a set of quantum rules (or gates) to her qubit depending on the 2 bits of information she wants to send. (See Table 5-3.)

Table 5-3. *Encoding Rules for Super Dense Coding*

Rules	Outcome States
00: I (identity gate)	I(00+11) = 00 + 11
01: X	X(00+11) = 10 + 01
10: Z	Z(00+11) = 00 − 11
11: ZX	ZX(00+11) = 10 − 11

3. Thus if she sends a 00, she does nothing to her qubit (applies the identity gate). If she sends a 01, then she applies the X gate (or bit flip). For a 10 she applies the Z gate. Note that the Z gate flips the sign (phase) of the qubit if the qubit is 1. Thus $Z\lvert 0\rangle = \lvert 0\rangle, Z\lvert 1\rangle = -\lvert 1\rangle$. Finally, if she sends 11, then she applies gates XZ to her qubit. Alice then sends her qubit to Bob for the final step in the process.

4. Bob receives Alice's qubit (qubit 0) and uses his qubit to reverse
   the process of the Bell state created by Eve. That is, he applies the
   CNOT gate to the first qubit followed by the Hadamard gate (H)
   and finally performs a measurement in both qubits to extract the 2
   classical bits encoded in Alice's qubit (see Table 5-4).

Table 5-4. *Qubit States After Recovery*

Gate	Outcome States	Details
CNOT	00 +10 11 + 01 00 − 10 11 − 10	We start with Alice's states from step 2: 00 + 11 10 + 01 00 − 11 10 − 11 The CNOT gate flips the second qubit if the first is 1 resulting in the states in column #2.
H	00 01 10 −11	Applying the Hadamard to the first qubit in the last row results in the outcomes in column #2. When Bob performs measurements in the computational basis states, he ends up with four possible outcomes with probability 1 each. These outcomes match what Alice meant to send in step 2 column #1. Note that the last outcome has a negative sign. Nevertheless, because the probability is calculated as the amplitude squared, the −1 becomes 1 which is correct.

Let's put all this together in a circuit within the IBM Q Experience Composer.

Circuit for Composer

Figure 5-6 shows the super dense coding circuit as well as the quantum assembly code within the Composer:

- The circuit begins by creating a Bell Pair; that is, it puts qubit[0] in
  superposition (using the Hadamard gate) and then entangles it with
  qubit[1] via the CNOT gate.

- The next two gates represent Alice's encoding rules. Remember that she applies the identity (nothing) to encode bits 00, X to encode 01, Z to encode 10, and ZX to encode 11. In this particular case, the encoded bits are 11. This is shown left of the barrier symbol in Figure 5-6. Note that the barrier will block execution until all gates are consumed by both qubits.

- To the right side of the barrier symbol, there is Bob's protocol. He basically does the reverse operation as Alice's. He applies the CNOT gate and then a Hadamard gate on the qubits. Finally a measurement is performed on both qubits to extract the 2 encoded classical bits.

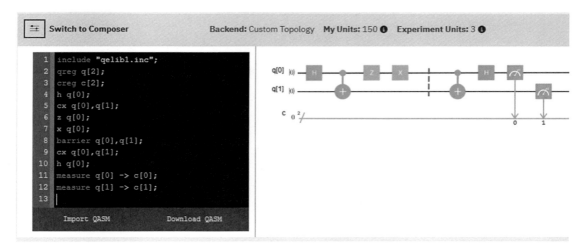

Figure 5-6. *Superdense circuit for Q Experience*

Run the preceding circuit in the simulator, and the result should be a bar graph with the probability for outcome 11 very close or equal to 1. This result should match the result obtained in the next section using a Python script.

Running Remotely Using Python

Listing 5-2 shows the equivalent Python script for the circuit in Figure 5-6:

- Lines 17–19 create 2 qubits and two classical registers to hold the outcomes.

- Next the *superdense* circuit is created with the entangled Bell Pair (lines 22–14).

- Alice encodes 11 by applying the ZX gates. Optionally, comment any of these statements to encode a different pair, and then make sure the result matches Alice's encoding scheme (lines 32–35).

- Bob reverses Alice's operation and measures the qubits (lines 38–41).

- Finally, the circuit gets executed in the remote simulator (ibmq_qasm_simulator) and the results displayed using Python's excellent plotting support.

Listing 5-2. Super Dense Coding Python Script

```python
import sys,time,math

# Importing QISKit
from qiskit import QuantumCircuit, QuantumProgram

sys.path.append('../Config/')
import Qconfig

# Import basic plotting tools
from qiskit.tools.visualization import plot_histogram

def main():
  # Quantum program setup
  Q_program = QuantumProgram()
  Q_program.register (Qconfig.APItoken, Qconfig.config["url"])

  # Creating registers
  q = Q_program.create_quantum_register("q", 2)
  c = Q_program.create_classical_register("c", 2)

  # Quantum circuit to make the shared entangled state
  superdense = Q_program.create_circuit("superdense", [q], [c])
  superdense.h(q[0])
  superdense.cx(q[0], q[1])
```

```
# For 00, do nothing
# For 10, apply X
# superdense.x(q[0])
# For 01, apply Z
# superdense.z(q[0])

# Alice: For 11, apply ZX
superdense.z(q[0])
superdense.x(q[0])
superdense.barrier()

# Bob
superdense.cx(q[0], q[1])
superdense.h(q[0])
superdense.measure(q[0], c[0])
superdense.measure(q[1], c[1])

circuits = ["superdense"]
print(Q_program.get_qasms(circuits)[0])

backend = "ibmq_qasm_simulator" #ibmqx2   quantum device
shots = 1024        # the number of shots in the experiment

result = Q_program.execute(circuits, backend=backend, shots=shots, max_
credits=3,  timeout=240)

print("Counts:" + str(result.get_counts("superdense")))

plot_histogram(result.get_counts("superdense"))


#############################################
# main
if __name__ == '__main__':
  start_time = time.time()
  main()
  print("--- %s seconds ---" % (time.time() - start_time))
```

Let's look at the results of a single run of Listing 5-2 in the next section.

Looking at the Results

The standard output of a run of Listing 5-2 is shown in the next paragraph:

```
C:\python36-64\python.exe p05-superdensecoding.py
OPENQASM 2.0;
include "qelib1.inc";
qreg q[2];
creg c[2];
h q[0];
cx q[0],q[1];
z q[0];
x q[0];
barrier q[0],q[1];
cx q[0],q[1];
h q[0];
measure q[0] -> c[0];
measure q[1] -> c[1];

Counts:{'11': 1024}
--- 167.52969431877136 seconds ---
```

The script dumps the assembly code of the circuit as well as the counts for the outcome: {'11': 1024} plus the execution time. The result count is used to calculate the probability of the outcome by dividing the number of shots (1024) by the outcome count (1024). Thus the probability is 1 for outcome 11, as shown in the plot run as the final step in Listing 5-2 (see Figure 5-7). Note that when executed in the simulator, the probability will always be 1, that is, counts = shots. However if you run in a real quantum device, because of noise and environmental error, the number of counts should be less the 1024 resulting in a probability less than 1.

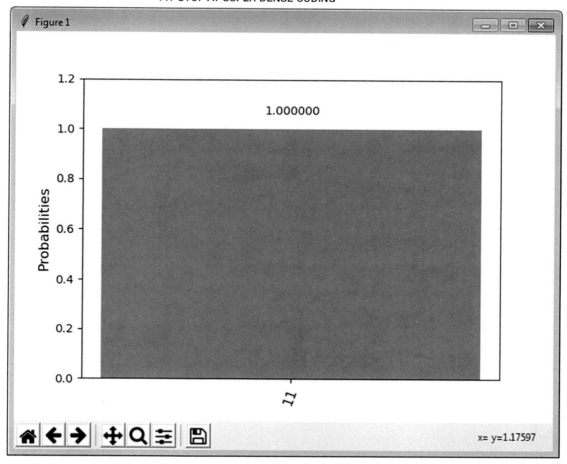

Figure 5-7. *Super dense coding plot result*

Thus super dense coding provides the means to encode 2 classical bits in a single qubit. Note that it is worth mentioning that quantum computation states that it is not possible to store more than a single classical bit per qubit which seems to contradict what has been shown in this protocol. As a matter of fact, there is no contradiction. The protocol works because Alice's and Bob's qubits are entangled via a Bell Pair. This allows for sending 2 classical bits in Alice's entangled qubit. All in all, you can store at most 2 classical bits per qubit provided that your qubit is entangled to another via a Bell Pair.

In general terms, this protocol could be interpreted as a set of modularized abstractions: a Bell Pair generator module to create 2 entangled qubits, followed by an information encoder module which applies Alice's rules to encode the 2 classical bits of information. Finally, a decoder module extracts the classical bits from the qubits provided by the Bell Pair as well as the encoder module (sort of a quantum zip/ unzip tool if you will). Super dense coding provides a high-level picture for quantum information processing and will help you understand the next item in this chapter: quantum teleportation.

Tip   This simple protocol was developed in 1992 by physicist Charles Bennett almost 70 years after the discovery of quantum mechanics. Despite its relative simplicity, it is not an obvious procedure, and remarkable things can be learned by studying it in depth.

Quantum Teleportation

Quantum teleportation is a procedure closely related to super dense coding. Perhaps the term teleportation is a little extravagant, as we are not really teleporting anything, at least not in the sci-fi/*Star Trek* sense. Formally quantum teleportation is the process by which the state of a qubit (ψ) can be transmitted from one location to another, with the help of classical communication and a Bell Pair discussed in the previous section. The procedure is summarized in Figure 5-8.

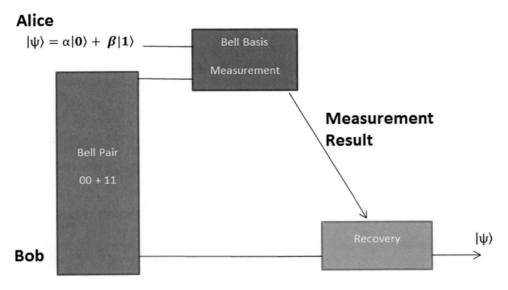

Figure 5-8. *Quantum teleportation workflow*

1. Alice and Bob start by sharing a Bell Pair of entangled qubits. One goes to Alice and the other goes to Bob at separate remote locations. Imagine that the Bell Pair is prepared by a third party (Eve).

2. Alice prepares her qubit to be teleported in state $|\psi\rangle = \alpha|0\rangle+\beta|1\rangle$. She then performs a Bell basis measurement of her qubit and the entangled qubit from the Bell Pair provided by Eve. Alice then sends the measurement result by classical means to Bob.

3. At this point there is a posterior state for Bob's qubit as a function of the measurement performed by Alice. This is the key to understanding the procedure; remember that both share an entangled qubit. Thus we'll see how Bob, by applying the appropriate quantum gate, can recover the original state ψ created by Alice.

Let's figure this out by looking at Bob's posterior state at the moment of Alice's measurement before the recovery operation. To do this, we write the joined states of the 3 qubits involved in the process. Note that the ket notation is ignored for simplicity. Thus given Alice's state $|\psi\rangle = \alpha|0\rangle+\beta|1\rangle$, if we combine it with the shared entangled qubit from the Bell Pair provided by Eve, we get

$$(\alpha 0 + \beta 1)\ (00 + 11)\ =\ \alpha 000 + \alpha 011 + \beta 100 + \beta 111 \qquad (1)$$

Now we need to write the state of the first 2 qubits using the Bell basis states

B0 = 00 + 11	00 = B0 − B1
B1 = 10 + 01	01 = B1 − B3
B2 = 00 − 11	10 = B1 − B3
B3 = 10 − 01	11 = B0 − B2

Expression (1) becomes

$$(\alpha 0 + \beta 1)\ (00 + 11)\ =\ B0\ (\alpha 0 + \beta 1) + B1\ (\alpha 1 + \beta 0) + B2\ (\alpha 0 - \beta 1) + B3$$
$$(-\alpha 1 + \beta 0) \qquad (2)$$

Expression (2) shows the states for the 3 qubits after Alice performs her measurement. Bob knows how to recover Alice's ψ by looking at the posterior state of the qubits in expression 2 (the states within the parenthesis). This is shown more clearly in Table 5-5.

Table 5-5. *Quantum Teleportation Recovery*

Bell State	Posterior State	Bob's Recovery Operation
B0	$\alpha 0 + \beta 1$	ψ
B1	$\alpha 1 + \beta 0$	$X\psi$
B2	$\alpha 0 - \beta 1$	$Z\psi$
B3	$-\alpha 1 + \beta 0$	$ZX\psi$

All in all, the quantum teleportation protocol provides the means to recover the state ψ of any qubit by sharing an entangled Bell Pair between two remote parties, hence the name teleportation. Now let's build a circuit for this protocol, run it in the simulator, and finally look at the results.

Circuit for Composer

Figure 5-9 shows the Composer circuit as well as the execution results (simulator only – no real device at this time) for the quantum teleportation protocol:

- The gates left of the barrier symbol (the dotted line) represent the Bell Pair prepared by the third party (Eve): qubits 1 and 2.

- Alice prepares her qubit (0) to a given state ψ. The actual value of ψ is irrelevant as it will be recovered by Bob at the final stage of the process. Alice receives qubit[1] from Eve, and qubit[2] goes to Bob.

- Alice performs a measurement on her qubits [0,1] (shown to the right of the dotted line) and sends the results by classical means to Bob.

- Bob applies the recovery rules to his qubit (2) mentioned in the previous section depending on the outcomes sent by Alice. Finally, after a measurement of qubit[2], Bob recovers the state ψ originally created by Alice. All this is made possible by the fact that Alice and Bob share an entangled pair of qubits which makes the whole thing work.

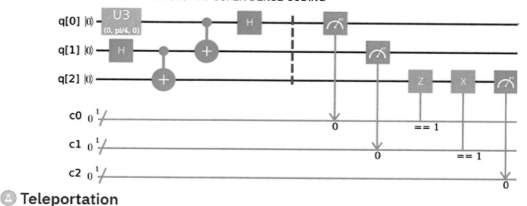

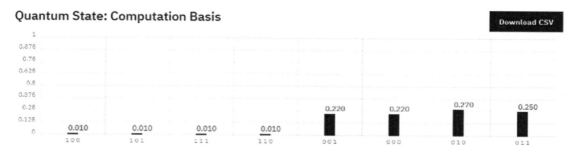

Figure 5-9. *Quantum teleportation circuit for the Composer*

Of course, the execution results in Figure 5-9 need to be massaged to verify that Bob's ψ matches Alice's. The best way to do this is to use a Python script. In the next section, we'll run the same circuit remotely and look at the results to verify the protocol works.

Running Remotely Using Python

In this section we use Python to run the quantum teleportation protocol remotely in the simulator. Note that, at this time, quantum teleportation cannot be run in a **real quantum device** on IBM Q Experience. This is due to the fact the hardware does not

213

support the rotation gate required by Alice to create her state ψ. Thus we'll use the remote simulator instead – a local Python simulator will be fine too. Listing 5-3 shows the protocol in action. In particular:

- Three qubits are created to be shared by both parties: Alice and Bob, plus three classical registers (c0, c1, c2) to store Alice's results (lines 20–23).

- The Bell Pair is prepared by Eve by applying a Hadamard gate (H) followed by a controlled NOT (CNOT) gate in qubits 1 and 2 (lines 35–37).

- Alice prepares her state ψ on qubit 0 by rotating on the Y-axis by π/4 radians (line 32).

- Alice now entangles her qubit[0] with the Bell Pair qubit given to her, qubit[1], to entangle them. She then performs a measurement in both and stores the outcomes in classical registers 0, 1 (lines 35–41).

- Now its Bob's turn: He applies a Z or X gate on his qubit (2) depending on the outcomes sent by Alice – if classical register 0 is 1, then he applies a Z gate. If classical register 1 is 1, then he applies an X gate. Then he measures his qubit and stores the outcome in classical register 2 (lines 47–50).

- The program is executed in the remote simulator (ibmq_qasm_simulator) and the results collected for display and verification (lines 58–79).

Tip   The source for this program is included in the book source under
`Workspace\Ch05\p05-teleport.py`.

Listing 5-3. Python Script for Quantum Teleportation

```
import sys,time,math
import numpy as np

# Importing QISKit
from qiskit import QuantumCircuit, QuantumProgram
```

```python
# Q Experience config
sys.path.append('../Config/')
import Qconfig

# Import basic plotting tools
from qiskit.tools.visualization import plot_histogram

def main():
    # Quantum program setup
    Q_program = QuantumProgram()
    Q_program. register(Qconfig.APItoken, Qconfig.config["url"])

    # Creating registers
    q = Q_program.create_quantum_register('q', 3)
    c0 = Q_program.create_classical_register('c0', 1)
    c1 = Q_program.create_classical_register('c1', 1)
    c2 = Q_program.create_classical_register('c2', 1)

    # Quantum circuit to make the shared entangled state (Bell Pair)
    teleport = Q_program.create_circuit('teleport', [q], [c0,c1,c2])
    teleport.h(q[1])
    teleport.cx(q[1], q[2])

    # Alice prepares her quantum state to be teleported,
    # psi = a|0> + b|1> where a = cos(theta/2), b = sin (theta/2), theta = pi/4
    teleport.ry(np.pi/4,q[0])

    # Alice applies CNOT to her two quantum states followed by H, to entangle
    them
    teleport.cx(q[0], q[1])
    teleport.h(q[0])
    teleport.barrier()

    # Alice measures her two quantum states:
    teleport.measure(q[0], c0[0])
    teleport.measure(q[1], c1[0])
```

```
circuits = ['teleport']
print(Q_program.get_qasms(circuits)[0])

##### BOB Depending on the results applies X or Z, or both, to his state
teleport.z(q[2]).c_if(c0, 1)
teleport.x(q[2]).c_if(c1, 1)

teleport.measure(q[2], c2[0])

# dump asm
circuits = ['teleport']
print(Q_program.get_qasms(circuits)[0])

# Execute inthe simulator (the real device does not support it yet)
#backend = "local_qasm_simulator"
backend = "ibmq_qasm_simulator"
shots = 1024        # the number of shots in the experiment

result = Q_program.execute(circuits, backend=backend, shots=shots
          , max_credits=3, timeout=240)

print("Counts:" + str(result.get_counts("teleport")))


# RESULTS
# Alice's measurement:
data = result.get_counts('teleport')
alice = {}
alice['00'] = data['0 0 0'] + data['1 0 0']
alice['10'] = data['0 1 0'] + data['1 1 0']
alice['01'] = data['0 0 1'] + data['1 0 1']
alice['11'] = data['0 1 1'] + data['1 1 1']
plot_histogram(alice)
```

```
#BOB
bob = {}
bob['0'] = data['0 0 0'] + data['0 1 0'] +  data['0 0 1'] + data['0 1 1']
bob['1'] = data['1 0 0'] + data['1 1 0'] +  data['1 0 1'] + data['1 1 1']
plot_histogram(bob)

############################################
# main
if __name__ == '__main__':
  start_time = time.time()
  main()
  print("--- %s seconds ---" % (time.time() - start_time))
```

To verify the results, the outcome counts returned by the simulator must be gathered for Alice and Bob. A plot of the results is the best way to verify that Alice's state ψ has been recovered by Bob. Here is a sample of what the simulator returns:

```
{'1 0 0': 37, '1 0 1': 45, '1 1 1': 43, '0 1 1': 215, '0 0 1': 200, '0 0
0': 206, '0 1 0': 230, '1 1 0': 48}
```

In this JSON string, the left side is the outcome(s) of the 3 qubits in reverse order. For example, in the first outcome 1 0 0: B(1) A(0) A(0) for Alice = A and Bob = B. To the right is the count obtained for that specific outcome. Remember that the probability of this outcome (used for graphing purposes) is calculated by dividing the outcome by the total number of shots (1024); thus

```
P(1 0 0) = 37/1024 = 0.036
```

The histogram plots for the results for Alice and Bob from the execution of Listing 5-3 are shown in Figure 5-10.

217

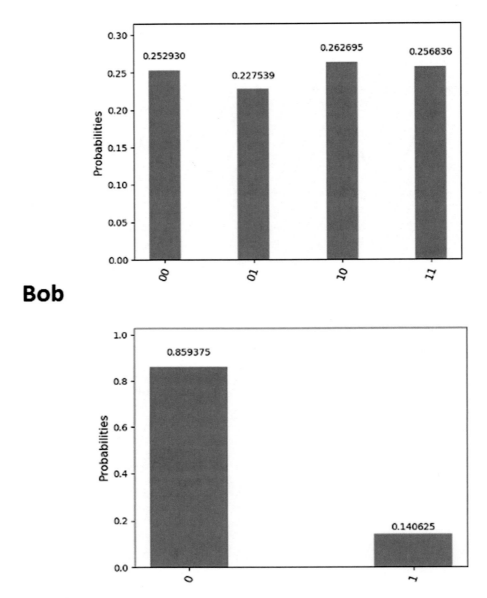

Figure 5-10. *Probability results for Alice and Bob measurements*

So what does this all mean? And how do we know that the state ψ has been recovered by Bob? Let's look at these results in more detail.

Looking at the Results

To interpret these results, first let's see how the probabilities are calculated from the counts retuned from Listing 5-3:

```
{'1 0 0': 37, '1 0 1': 45, '1 1 1': 43, '0 1 1': 215, '0 0 1': 200, '0 0
0': 206, '0 1 0': 230, '1 1 0': 48}
```

Using these counts we can calculate the probabilities for Alice's and Bob's outcomes shown in Figure 5-10 (see Table 5-6).

Table 5-6. *Probability Results for the Quantum Teleportation Experiment*

Row			Outcome	Count	Probability	Alice	Probability Sum
0	Alice(00)	Bob(0)	0 0 0	206	0.201171875	0 0	0.237304688
1	Alice(01)	Bob(0)	0 0 1	200	0.1953125	1 0	0.239257813
2	Alice(10)	Bob(0)	0 1 0	230	0.224609375	0 1	0.271484375
3	Alice(11)	Bob(0)	0 1 1	215	0.209960938	1 1	0.251953125
4	Alice(00)	Bob(1)	1 0 0	37	0.036132813		
5	Alice(01)	Bob(1)	1 0 1	45	0.043945313	**Bob**	
6	Alice(10)	Bob(1)	1 1 0	48	0.046875	0	0.831054688
7	Alice(11)	Bob(1)	1 1 1	43	0.041992188	1	0.168945313

As shown in Table 5-6, to calculate the total probability of Alice's outcome 00, we need to sum the probability columns for rows 0 and 4. That is, P(A00) = 0.201 + 0.036 = 0.237. The same rules apply to Bob. For example, P(B0) = 0.20 + 0.19 + 0.22 + 0.20 = 0.83 (add probability columns for rows 0-3). This is shown on the right side for all outcomes of Alice and Bob. This is how the script in Listing 5-3 massages the data before plotting the results shown in Figure 5-10. But what does this mean, and how do we know that Bob has recovered Alice's ψ? Let's look at Bob's total probability for his qubit:

Bob

0   `0.20 + 0.19 + 0.22 + 0.20 = 0.83`

1   `0.036 + 0.043 + 0.046 + 0.041 = 0.168`

Quantum mechanics says that the probability of ψ is given by $P(\psi) = |\psi|^2$. That is, the probability density is the modulus squared of ψ. Now remember that Alice prepared ψ as

$$\Psi = RY(\theta)\, Where\, \theta = \frac{\pi}{4}$$

That is, Alice applied a π/4 rotation over the Y-axis on her qubit. To see this more clearly, let's visualize the state ψ using geometry (see Figure 5-11):

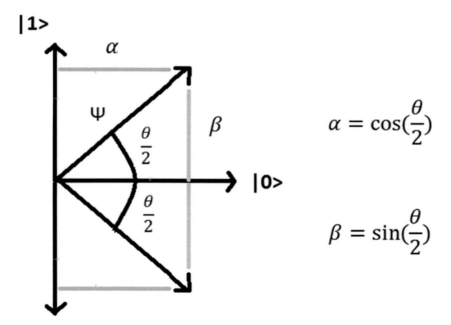

Figure 5-11. *Superimposed state for Alice's Ψ*

Remember that the superimposed state ψ is described in terms of the complex coefficients α and β as

$$\Psi = \alpha|0\rangle + \beta|1\rangle$$

$$Probability\,|0\rangle = |\alpha|^2, Probability\,|1\rangle = |\beta|^2$$

But from Figure 5-11, we can represent the coefficients as α = cos(θ/2) and β = sin(θ/2). Thus, finally, if θ= π/4, then

$$Probability\,(\alpha) = |\cos(\pi/8)|^2 = 0.85$$

$$Probability\,(\beta) = |\sin(\pi/8)|^2 = 0.14$$

This matches Bob's results from the plot created by the teleportation Listing 5-3 (see Figure 5-12). Great success!

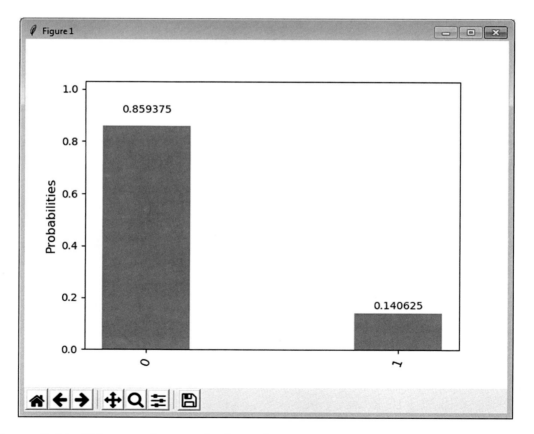

Figure 5-12. *Teleportation results for Bob*

You have taken the first step to understand the remarkable information processing capabilities of quantum systems. We started with a simple procedure to exploit the source of true randomness intrinsic to quantum mechanics to generate random numbers. We also explored two bizarre protocols: super dense coding to encode classical information and quantum teleportation to recover the state of a qubit by a remote party. These protocols have been described using circuits for the IBM Q Experience as well as Python scripts for remote execution in a simulator or real quantum device. Results have been collected and explained to understand what goes on behind the scenes. The next chapter explores the lighter side of quantum computing, by having fun creating a simple game using quantum gates: a needed break before we get to heavy stuff in later chapters.

CHAPTER 6

Fun with Quantum Games

In this chapter you will learn how to implement a basic game in a quantum computer. For this purpose we use the quintessential Quantum Battleship distributed with the QISKit Python tutorial. The first part looks at the mechanics of the game including

- Using qubits to represent ship positions in the board

- How to calculate damage percentages using a quantum program to be run in the local, remote simulator or real quantum device

- How to perform rotations on the X-axis of a single qubit using a partial NOT quantum gate

Yet we don't stop there. The second part of this chapter takes things to the next level. The game is given a major face lift by showing you how to implement a cloud-based Quantum Battleship with the following features:

- A browser-based user interface with interactive boards to place ship or bombs. The game mechanics remain the same nonetheless.

- A CGI-based script using the Apache Web Server to consume game events and dispatch them to the quantum program.

- A modified version of the original quantum program to perform partial NOT rotations on the qubits to calculate ship damage. Most of the original code remains intact.

You will learn how the original code can be modularized and reused for a different take of Quantum Battleship in the cloud. Let's get started.

© Vladimir Silva 2018
V. Silva, *Practical Quantum Computing for Developers*, https://doi.org/10.1007/978-1-4842-4218-6_6

Quantum Battleship with a Twist

In this section we look at a game distributed with the QISKit tutorial called *Quantum Battleship*. The program uses 5 qubits to represent a board where each player places three ships. It then asks each player to place a bomb in a position 0-5. Finally, damage for each ship is calculated by a quantum program that uses a two-pulse single-qubit gate: *U3(theta, phi, lambda)*. This gate is called a partial NOT gate and performs rotation on axes X, Y, Z by theta, phi, or lambda radians.

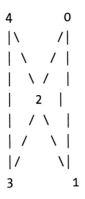

```
4        0
|\      /|
| \    / |
|  \  /  |
|   2   |
|  / \  |
| /   \ |
|/     \|
3        1
```

In this particular case, ship damage is calculated by doing a series of partial rotations on the X-axis (theta) using the number of bombs for that position. If the damage for a position (or ship) exceeds 95%, the ship is destroyed, and once the entire player's fleet is smashed, a winner is declared and the game is over. This is just the standard battleship game we all played as children but using a quantum computer or simulator in the background.

Note   The game was written by James Wootton from the University of Basel and contributed to the QISKit Python tutorial. A modified version of the original code by Wootton is available from the source of this book at `Workspace\Ch06\battleship\BattleShip.py` (minus some unnecessary fancy text).

Let's run the program and take a look at the game mechanics.

Setup Instructions

From the book source, execute the program *BattleShip.py* as described in the following paragraphs:

- For CentOS 6 or 7 or any Fedora-like OS, activate your Python virtual environment. This is required only if you have multiple versions of Python, for example, 2.7 and 3.6. Remember that you must use 3.5 or later. Instructions on how to set up a virtual Python environment were covered in Chapter 3.

- Copy the script Workspace/Ch06/battleship/BattleShip.py and the configuration file *Qconfig.py* from the book source to your workspace and execute it (as shown in the next fragment).

```
# Activate Python3 virtual environment at $HOME/qiskit/qiskit
$ source $HOME/qiskit/qiskit/bin/activate
$ python BattleShip.py
############### Quantum Battle Ship ################
Do you want to play on the real device? (y/n) n
```

Let's look at how the program works.

Initialization

Listing 6-1 shows the script initialization. It starts by doing basic Python tasks:

- It loads system libraries: sys and QuantumProgram required for all QISKit operations.

- It makes sure you are using Python 3.5 or later.

- It asks if you wish to use a simulator or a real quantum computer. It then sets the number of shots for the execution to the default 1024.

Listing 6-1. Script Initialization

```
###################################
# Quantum Battleship from tutorial @
# https://github.com/QISKit/qiskit-tutorial
###################################
import sys

# Checking the version of PYTHON; we only support > 3.5
if sys.version_info < (3,5):
    raise Exception('Please use Python version 3.5 or greater.')

from qiskit import QuantumProgram
import Qconfig
import getpass, random, numpy, math

## 1. Select a backend: IBM simulator (ibmqx_qasm_simulator) or real chip
ibmqx2

d = input("Do you want to play on the real device? (y/n)\n").upper()
if (d=="Y"):
    device = 'ibmqx2'
else:
    device = 'ibmqx_qasm_simulator'

# note that device should be 'ibmqx_qasm_simulator', 'ibmqx2' or 'local_
qasm_simulator'
# while we are at it, let's set the number of shots
shots = 1024
```

Tip   To run a quantum program on a real device, you must place the configuration file (Qconfig.py) in the same location as the main script. The configuration contains your required API token and IBM Q Experience end point.

```
APItoken = 'YOUR API TOKEN'
config = {
    'url': 'https://quantumexperience.ng.bluemix.net/api',
}
```

Now let's place some ships in the board.

Set Ships in the Board

The program uses a rudimentary text-based interface for all user input. Listing 6-2 shows the logic to enter ships for each player. Press ENTER to start, and type the position for up to three ships per player (the positions are zero based).

- The script can bypass user choice and select random positions, or else the player must enter positions for three ships.

- Positions are stored in the two-dimensional list shipPos where shipPos[0] contains the positions for player 1 and shipPos[1] contains positions for player 2. Remember that only three ships are allowed per player.

Listing 6-2. Setting Ships on the Board

```
######## 2. players to set up their boards.
randPlace = input("> Press Enter to start placing ships...\n").upper()

# The variable ship[X][Y] will hold the position of the Yth ship of player
X+1
shipPos = [ [-1]*3 for _ in range(2)]

# loop over both players and all three ships for each
for player in [0,1]:

    # if we chose to bypass player choice and do random, we do that
    if ((randPlace=="r")|(randPlace=="R")):
        randPos = random.sample(range(5), 3)
        for ship in [0,1,2]:
            shipPos[player][ship] = randPos[ship]
```

```
    else:
        for ship in [0,1,2]:

            # ask for a position for each ship,
            choosing = True
            while (choosing):

                # get player input
                position = getpass.getpass("Player " + str(player+1)
                    + ", choose a position for ship " + str(ship+1) +
                    " (0-4)\n" )

                # see if the valid input and ask for another if not
                if position.isdigit(): # valid answers  have to be integers
                    position = int(position)
            # they need to be between 0 and 5
                    if (position in [0,1,2,3,4]) and (not position in
                    shipPos[player]):
                        shipPos[player][ship] = position
                        choosing = False
                        print ("\n")
                    elif position in shipPos[player]:
                        print("\nYou already have a ship there. Try
                        again.\n")
                    else:
                        print("\nThat's not a valid position. Try again.\n")
                else:
                    print("\nThat's not a valid position. Try again.\n")
```

The following section shows the standard output, very rudimentary but so far so good.

```
Do you want to play on the real device? (y/n)
n
Player 1, choose a position for ship 1 (0, 1, 2, 3 or 4)
0
Player 1, choose a position for ship 2 (0, 1, 2, 3 or 4)
1
```

```
Player 1, choose a position for ship 3 (0, 1, 2, 3 or 4)
2

Player 2, choose a position for ship 1 (0, 1, 2, 3 or 4)
0

Player 2, choose a position for ship 2 (0, 1, 2, 3 or 4)
1

Player 2, choose a position for ship 3 (0, 1, 2, 3 or 4)
2
```

The interesting stuff occurs in the main loop. Let's take a look.

Main Loop and Results

The main loop performs the following tasks:

- It asks both players to place one bomb in position [0-4]. A count of the bomb is stored in a two-dimensional list of five elements (two players, five bomb counts). Note that the player can bomb the same position multiple times; thus if player 1 bombs position 0 twice, then bombs = [[2,0,0,0,0],[0,0,0,0,0]].

- It creates a QuantumProgram to hold 5 qubits (1 per position in the board) and five classical registers to hold the measurement results.

- If a bomb position matches the opposing player's ship position (from the shipPos list), the damage is calculated by performing one rotation over the X-axis per bomb count using a single-qubit partial NOT gate: gridScript.u3(1/(ship +1) * math.pi, 0.0, 0.0, q[position]). Note that the effectiveness of the bomb also depends on which ship is bombed (0, 1, 2).

- To complete the circuit, a measurement is performed in the qubit for the position and the result stored in the respective classical register: gridScript.measure(q[position], c[position]).

- Next, the program is executed in the target device, and the results stored in the two-dimensional list grid. For example, if position 0 for player 1 is bombed, then grid = [[1,0,0,0,0],[0,0,0,0,0]]. The following paragraph shows how this is done:

```
results = Q_program.execute(["gridScript"], backend=device,
shots=shots)
grid[player] = results.get_counts("gridScript")
```

- The results are checked for errors. If no errors, then a damage percentage between [0, 1] is calculated if the grid list contains a 1 for that position. The percentages are kept in the two-dimensional list damage. Thus damage [[0.95, 0, 0, 0, 0], [0, 0, 0, 0, 0]] indicates that player 1 ship in position 0 has been destroyed.

- It finally presents the results to the players in a simple text-based interface. The process repeats itself until all ships are destroyed and a winner is declared (see Listing 6-3).

Listing 6-3. Battleship Main Loop

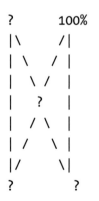

```
?          100%
|\        /|
| \      / |
|  \  /  |
|    ?    |
|  / \   |
| /     \ |
|/        \|
?               ?
```

```
########### 3. Main loop.
# Each iteration starts by asking players where on the opposing grid they
want a bomb.
# The quantum computer calculates the effects of the bombing, and the
results are presented.
# The game continues until all the ships of one player are destroyed.
game = True
```

```
# the variable bombs[X][Y] holds the number of times position Y has been
bombed by player X+1
bomb = [ [0]*5 for _ in range(2)] # all values are initialized to zero

# the variable grid[player] will hold the results for the grid of each
player
grid = [{},{}]

while (game):

    input("> Press Enter to place some bombs...\n")

    # ask both players where they want to bomb
    for player in range(2):

        print("\n\nIt's now Player " + str(player+1) + "'s turn.\n")

        # keep asking until a valid answer is given
        choosing = True
        while (choosing):

            # get player input
            position = input("Choose a position to bomb (0, 1, 2, 3
            or 4)\n")

            # see if this is a valid input. ask for another if not
            if position.isdigit(): # valid answers  have to be integers
                position = int(position)
                if position in range(5):
                    bomb[player][position] = bomb[player][position] + 1
                    choosing = False
                    print ("\n")
                else:
                    print("\nThat's not a valid position. Try again.\n")
            else:
                print("\nThat's not a valid position. Try again.\n")

    # now we create and run the quantum program for each player
    for player in range(2):
```

```
    if device=='ibmqx2':
        print("\nUsing a quantum computer for Player " + str(player+1)
        + "'s ships.\n")
    else:
        print("\nUsing the simulator for Player " + str(player+1) + "'s
        ships.\n")

    # now to set up the quantum program (QASM) to simulate the grid for
    this player

    Q_program = QuantumProgram()
    # set the APIToken and API url
    Q_program.set_api(Qconfig.APItoken, Qconfig.config["url"])
    # declare register of 5 qubits
    q = Q_program.create_quantum_register("q", 5)
    # declare register of 5 classical bits to hold measurement results
    c = Q_program.create_classical_register("c", 5)
    # create circuit
    gridScript = Q_program.create_circuit("gridScript", [q], [c])

    # add the bombs (of the opposing player)
    for position in range(5):
        # add as many bombs as have been placed at this position
        for n in range( bomb[(player+1)%2][position] ):
            # the effectiveness of the bomb
            # (which means the quantum operation we apply)
            # depends on which ship it is
            for ship in [0,1,2]:
                if ( position == shipPos[player][ship] ):
                    frac = 1/(ship+1)
                    # add this fraction of a NOT to the QASM
                    gridScript.u3(frac * math.pi, 0.0, 0.0,
                    q[position])

    #finally, measure them
    for position in range(5):
        gridScript.measure(q[position], c[position])
```

```
    # to see what the quantum computer is asked to do, we can print the
    QASM file
    # this lines is typically commented out
    #print( Q_program.get_qasm("gridScript") )

    # compile and run the QASM
    results = Q_program.execute(["gridScript"], backend=device,
    shots=shots)

    # extract data
    grid[player] = results.get_counts("gridScript")

# we can check up on the data if we want
# these lines are typically commented out
#print( grid[0] )
#print( grid[1] )

# if one of the runs failed, tell the players and start the round again
if ( ( 'Error' in grid[0].values() ) or ( 'Error' in grid[1].
values() ) ):

    print("\nThe process timed out. Try this round again.\n")

else:

    # look at the damage on all qubits (we'll even do ones with no
    ships)
    # # this will hold the prob of a 1 for each qubit for each player
    damage = [ [0]*5 for _ in range(2)]

    # for this we loop over all 5 bit strings for each player
    for player in range(2):
        for bitString in grid[player].keys():
            # and then over all positions
            for position in range(5):
                # if the string has a 1 at that position, we add a
                contribution to the damage
                # remember that the bit for position 0 is the rightmost
                one, and so at bitString[4]
```

```
                if (bitString[4-position]=="1"):
                    damage[player][position] += grid[player]
                    [bitString]/shots

    # give results to players
    for player in [0,1]:

        input("\nPress Enter to see the results for Player
        " + str(player+1) + "'s ships...\n")

        # report damage for qubits that are ships, with significant
        damage
        # ideally this would be non-zero damage,
        # so we choose 5% as the threshold
        display = [" ? "]*5
        # loop over all qubits that are ships
        for position in shipPos[player]:
            # if the damage is high enough, display the damage
            if ( damage[player][position] > 0.1 ):
                if (damage[player][position]>0.9):
                    display[position] = "100%"
                else:
                    display[position] = str(int( 100*damage[player]
                    [position] )) + "% "

        print("Here is the percentage damage for ships that have been
        bombed.\n")
        print(display[ 4 ] + "     " + display[ 0 ])
        print(" |\      /|")
        print(" | \    / |")
        print(" |  \ /   |")
        print(" |   " + display[ 2 ] + " |")
        print(" |  / \   |")
        print(" | /    \ |")
        print(" |/      \|")
        print(display[ 3 ] + "     " + display[ 1 ])
        print("\n")
```

```
print("Ships with 95% damage or more have been destroyed\n")

print("\n")

# if a player has all their ships destroyed, the game is over
# ideally this would mean 100% damage, but we go for 90%
because of noise again
if (damage[player][ shipPos[player][0] ]>.9) and
(damage[player][ shipPos[player][1] ]>.9)
  and (damage[player][ shipPos[player][2] ]>.9):
    print ("***All Player " + str(player+1) + "'s ships have
    been destroyed!***\n\n")
    game = False

if (game is False):
    print("")
    print("=======GAME OVER=======")
    print("")
```

Note that if the damage exceeds 90%, the ship is marked as destroyed. Listing 6-4 shows the results of one game interaction.

Listing 6-4. Game Standard Output for One Game Interaction

```
> Press Enter to place some bombs...

It's now Player 1's turn.
Choose a position to bomb (0, 1, 2, 3 or 4)
0
It's now Player 2's turn.
Choose a position to bomb (0, 1, 2, 3 or 4)
0

We'll now get the simulator to see what happens to Player 1's ships.
We'll now get the simulator to see what happens to Player 2's ships.

Press Enter to see the results for Player 1's ships...
Here is the percentage damage for ships that have been bombed.
```

```
?         100%
|\       /|
| \     / |
|  \ /   |
|   ?    |
|  / \   |
| /   \  |
|/     \|
?         ?
```

Ships with 95% damage or more have been destroyed

Press Enter to see the results for Player 2's ships...
Here is the percentage damage for ships that have been bombed.

```
?         100%
|\       /|
| \     / |
|  \ /   |
|   ?    |
|  / \   |
| /   \  |
|/     \|
?         ?
```

Ships with 95% damage or more have been destroyed

Thus the main loop continues until a winner is declared. All in all you have learned how a simple game can be implemented to make use of a quantum computer to perform simple damage calculations via rotations in the X-axis of a qubit. This version is rudimentary but interesting nonetheless. However we can do better; in the next section, we give this game an improved look and feel.

Cloud Battleship: Modifying for Remote Access

It is really cool being able to play Battleship in quantum computer via a simple text interface, but it is much cooler playing the same game on a web browser in the cloud. In this section we modify the Quantum Battleship and give it a much-needed face lift (see Figure 6-1).

236

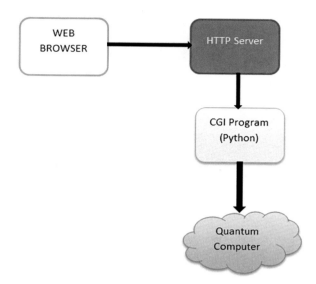

Figure 6-1. *Layout for Quantum Battleship in the cloud*

The idea is to

- Ditch the boring text-based interface in favor of an HTML page that can be deployed in the cloud.

- Use the Apache HTTP Server Common Gateway Interface (CGI) to deploy the quantum logic in a script using Python's excellent CGI support.

- Let the player select the device where to run the calculations: local simulator, remote simulator, or real quantum device.

Let's do this in a series of exercises described in the next sections.

Exercise 1: Decouple the User Interface from the Game Logic

A basic principle of object-oriented design: Never mix the presentation (user interface) with the business logic. This is so that modularized components can be built and reused all over. It saves a ton of time and money. In the case of Battleship, we need to remove or comment

- The first section of the script which reads the position of the ships for each player (a good chunk of code), being careful not to remove any data structures or variables.

- All print statements and keyboard input statements.

- The main *while* loop of the game that keeps asking for a position to bomb must be removed also. The script must terminate after it consumes data from the HTTP request. It cannot have infinite loops or else the request will hang.

- Add Python's CGI support to the script so the data can be read from the HTTP request including

  - Ship positions for each player

  - Bomb positions and count for each player

  - Device where to run the quantum computations

- The script must return a damage report (preferably in JSON) via the HTTP response for the browser to render in Javascript.

Note that most of the code will be reused: data structures, local variables, and quantum logic; it is just a matter of commenting all input and print statements. The solution to this (and each) exercise is shown at the end of this section.

Exercise 2: Build a Web Interface for the Ship-Bomb Boards

Build an HTML graphical user interface similar to the text-based UI, and use AJAX to send requests asynchronously to the CGI script. Get the damage results back and finally update the player boards. The improved looks are shown in Figure 6-2.

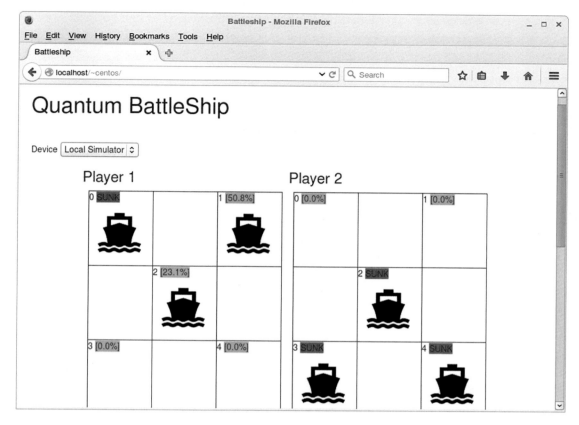

Figure 6-2. *User interface for the new Quantum Battleship*

- The HTML file will have four 3x3 boards. The top boards will be used to place three ships in five qubit locations. These will be implemented as HTML checkboxes (<INPUT TYPE="checkbox">). We will use CSS to replace the toggle button with an image, so when the player clicks the box, the ship image will be toggled instead.

- The bottom boards will allow the players to place bombs in five locations, using the same CSS in the preceding paragraph, but they will be implemented as <INPUT TYPE="radio"> so that multiple bombs can be placed per location.

- Even though the boards are 3x3, only five locations are allowed for user input corresponding to each qubit in the quantum program.

239

- Each ship location will display a qubit number and a colorized damage percentage returned by the backend.

- The game mechanics are exactly the same as the text-based version. Each player places three ships in the board, and then each takes a turn placing a bomb and clicks Submit. The Python CGI script will receive the request via AJAX, run the quantum program created in exercise 1, and return a damage result to be rendered in Javascript.

- Note that all game state, arrays, variables, and other data, is kept in the client HTML; therefore we must use AJAX to send the request **asynchronously**, or else the data will be lost every time a player submits. There will be no page refreshes.

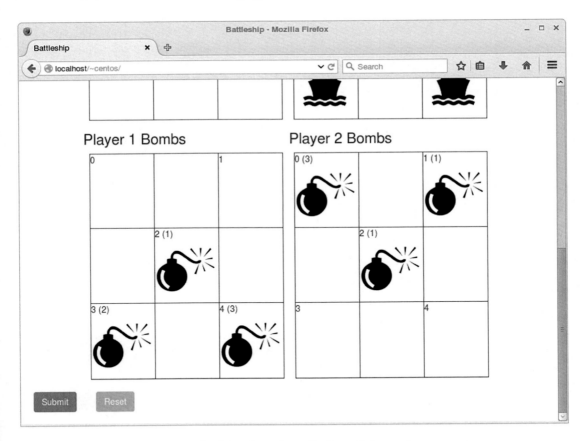

Figure 6-3. *Quantum Battleship showing the bomb boards*

Figure 6-3 shows the bottom 3x3 boards displaying a qubit number, a click count per bomb, and the image radio buttons rendered using CSS. Once each player places three ships and selects a location to bomb, the *Submit* button is clicked to send the AJAX request to the server. A *Reset* button is also available to restart the game at any point. Note that all state is kept in the client (browser). No data will be kept in the Python script as HTTP is a stateless request-response protocol. This means that when a request is received, the program is executed by the Web Server, a response is printed to the request output buffer, and the program terminates. As with the previous exercise, the solution is at the end of this section.

Exercise 3: Deploy and Troubleshoot in Apache HTTPD

Once all the pieces are in place, it's time to deploy to the Web Server. I will use Apache HTTPD in CentOS 6, but this should work for any CentOS, Fedora, or Red Hat flavor (probably for any current Linux distribution with Apache HTTPD). Note that each flavor has its own idiosyncrasies when it comes to configuring system software. For example, CentOS focuses in stability and security which gave me a lot of headaches configuring HTTPD and Python.

Solution 1: A Reusable Python Program

This section presents the Python CGI script that receives the HTTPD request from the browser and replies with a JSON string containing the damage report and other information as well. The first part of the program remains the same except that the input now must be parsed from the HTTP request using Python's *cgi* library (see Listing 6-5).

Listing 6-5. Modularized Quantum Battleship Initialization

```
import sys
from qiskit import QuantumProgram
import Qconfig
import getpass, random, numpy, math

import cgi
import cgitb
```

```
# solve the relative dependencies if you clone QISKit from the Git repo and
use like a global.
sys.path.append('../../qiskit-sdk-py/')

# debug
cgitb.enable(display=0, logdir=".")

# The variable ship[X][Y] will hold the position of the Yth ship of player
X+1
# all values are initialized to the impossible position -1|
shipPos = [ [-1]*3 for _ in range(2)]

# the variable bombs[X][Y] will hold the number of times position Y has
been bombed by player X+1
bomb = [ [0]*5 for _ in range(2)] # all values are initialized to zero
```

Listing 6-5 shows the first section of the script. Lines 6 and 7 import the Python libraries: **cgi** and **cgitb** (CGI Toolbox) used to read from the HTTP request and debug the CGI program, respectively.

Tip   The lines in the following paragraph activate a special exception handler that will display detailed reports in the web browser if any error occurs.

```
import cgitb
cgitb.enable()
```

Keep in mind that, if an error occurs, we cannot show the guts of the program as the client expects a response in JSON format, so we must save any error report to the current working directory instead, with code like this:

```
cgitb.enable(display=0, logdir=".")
```

The preceding code will save a lot of headaches during development as any exception will be dumped into a neat HTML document in the current working directory. The format of this document is shown in the "Troubleshooting" section of this chapter. Listing 6-5 also shows the data structures used to store the game state. These are the same as the old version:

- *shipPos*: A two-dimensional list that stores the positions for three ships per player initialized to -1; thus shipPos = [[-1, -1, -1], [-1, -1, -1]].

- *bomb*: A two-dimensional list that stores bomb counts per position per player initialized to 0: bomb = [[0,0,0,0,0], [0,0,0,0,0]]. Note that the same position can be bombed multiple times; therefore the need to store counts. This list will be used to calculate ship damage.

Next, the script reads the game data from the HTTP request (see Listing 6-6).

Listing 6-6. Reading Data from the HTTP Request

```
# CGI - parse HTTP request
form = cgi.FieldStorage()

ships1 = form["ships1"].value
ships2 = form["ships2"].value
bombs1 = form["bombs1"].value
bombs2 = form["bombs2"].value

# 'local_qasm_simulator', 'ibmqx_qasm_simulator'
device = str(form["device"].value)

shipPos[0] = list(map(int, ships1.split(","))) # [0,1,2]
shipPos[1] = list(map(int, ships2.split(","))) # [0,1,2]

bomb[0] = list(map(int, bombs1.split(",")))
bomb[1] = list(map(int, bombs2.split(",")))

stdout = "Ship Pos: " + str(shipPos) +  " Bomb counts: " + str(bomb) + "<br>"
```

- To read data from the HTTP request, use form = cgi. FieldStorage(). This CGI call returns a dictionary or hash map of key-value pairs used to extract query string parameters from the request. In this particular case, the values expected are

  - *ships1*: A three-element JSON array of player 1 ship positions.

  - *ships2*: A three-element JSON array of player 2 ship positions.

- *bombs1*: A five-element JSON array of bomb counts for player 1.

- *bombs2*: A five-element JSON array of bomb counts for player 2.

- *device*: The device where the quantum program will be executed. This can be

- *local_qasm_simulator*: Local simulator packed with the QISKit

- *ibmq_qasm_simulator*: Remote simulator provided by IBM

- *ibmqx2*: 5-qubit quantum processor provided by IBM Q Experience

- The great thing about Python is that the JSON provided by the HTTP request can be mapped to its excellent collection support in a snap:

```
shipPos[0] = list(map(int, ships1.split(",")))
bomb[0] = list(map(int, bombs1.split(",")))
```

Tip In Python, the `split(SEPARATOR)` system call is used to create a list of elements of type *String*. But we need a list of integers instead. For that purpose we use the `map(DATA-TYPE, LIST)` system call. Note that in Python 3, map returns a hash map (dictionary); therefore we must use the `list` system call to convert to a list of integers we need. This is great because it allows the script to reuse the old data structures and keep most of the quantum logic intact.

The last line of Listing 6-6 is simply a string buffer of standard output that will be returned to the browser for debugging purposes. Finally Listing 6-7 shows the guts of the script which remain mostly intact.

Listing 6-7. Quantum Script Main Section

```
# the variable grid[player] will hold the results for the grid of each
player
grid = [{},{}]

# now we create and run the quantum programs that implement this on the
grid for each player
```

```
for player in range(2):

    # now to set up the quantum program (QASM) to simulate the grid for
    this player

    Q_program = QuantumProgram()
    Q_program.set_api(Qconfig.APItoken, Qconfig.config["url"])

    # declare register of 5 qubits
    q = Q_program.create_quantum_register("q", 5)
    # declare register of 5 classical bits to hold measurement results
    c = Q_program.create_classical_register("c", 5)
    # create circuit
    gridScript = Q_program.create_circuit("gridScript", [q], [c])

    # add the bombs (of the opposing player)
    for position in range(5):
        # add as many bombs as have been placed at this position
        for n in range( bomb[(player+1)%2][position] ):
            # the effectiveness of the bomb
            # (which means the quantum operation we apply)
            # depends on which ship it is
            for ship in [0,1,2]:
                if ( position == shipPos[player][ship] ):
                    frac = 1/(ship+1)
                    # add this fraction of a NOT to the QASM
                    gridScript.u3(frac * math.pi, 0.0, 0.0, q[position])

    #finally, measure them
    for position in range(5):
        gridScript.measure(q[position], c[position])

    # to see what the quantum computer is asked to do, we can print the
    QASM file
    # this lines is typically commented out
    #print( Q_program.get_qasm("gridScript") )

    # compile and run the QASM
```

```
    results = Q_program.execute(["gridScript"], backend=device,
    shots=shots)

    # extract data
    grid[player] = results.get_counts("gridScript")
# if one of the runs failed, tell the players and start the round again
if ( ( 'Error' in grid[0].values() ) or ( 'Error' in grid[1].values() ) ):

    stdout += "The process timed out. Try this round again.<br>"

else:

    # look at the damage on all qubits (we'll even do ones with no ships)
    damage = [ [0]*5 for _ in range(2)]

    # for this we loop over all 5 bit strings for each player
    for player in range(2):
        for bitString in grid[player].keys():
            # and then over all positions
            for position in range(5):
                # if the string has a 1 at that position, we add a
                contribution to the damage
                # remember that the bit for position 0 is the rightmost
                one, and so at bitString[4]
                if (bitString[4-position]=="1"):
                    damage[player][position] += grid[player][bitString]/
                    shots

    stdout += "Damage: " + str(damage) + "<br>"
```

A few minor changes have been made to the main section of the original script:

- All print statements have been removed. Instead, a standard output
  string buffer is used to return information back to the client. This is
  done because Python's print will dump information directly into the
  HTTP response which will mess up the JSON format we must return
  back (Javascript expects proper JSON from AJAX). Note that this is a
  purely optional but helpful step meant to return debug information

back to the client. All in all, you can get away by simply commenting all print statements (of course, if an error occurs, you'll have a hard time figuring out what went wrong).

- All user input statements (read bomb position, *Press enter to continue*, and others) have been removed. Remember that ship positions and bomb counts are mapped from the HTTP request.

- The original script uses an endless while loop to read bomb positions. This loop has been removed, or else the script will run forever and hang the HTTP request.

Finally, the script returns a JSON document back to the browser for UI updates as shown in Listing 6-8.

Listing 6-8. Sending the Response Back to the Browser

```
# Response
print ("Content-type: application/json\n\n")
print ("{\"status\": 200, \"message\": \"" + stdout + "\", \"damage\":" +
str(damage) + "}")
```

To send a response back to the browser using Python CGI, simply print the standard HTTP response to standard output. That is, one or more HTTP headers followed by two line feeds and the response body. For example, to send a JSON document for the damage, we use the fragment:

```
Content-type: application/json
{ "status" : 200, "message": "Device ibmqx_qasm_simulator", "damage":
[[0.5, 0, 0, 0, 0], [0, 0.9, 0, 0, 0]]}
```

The preceding JSON document indicates damage for player 1 qubit(0) and player 2 qubit(1). This document will be parsed by the browser AJAX code and the values updated on screen.

Tip   The code for this exercise is available from the book source at `Workspace\Ch06\battleship\cgi-bin\qbattleship.py`.

Solution 2: User Interface

The web page uses a 2x2 HTML table to render four 3x3 inner tables representing the ship and bomb boards for each player as shown in Figure 6-4 and Listing 6-9.

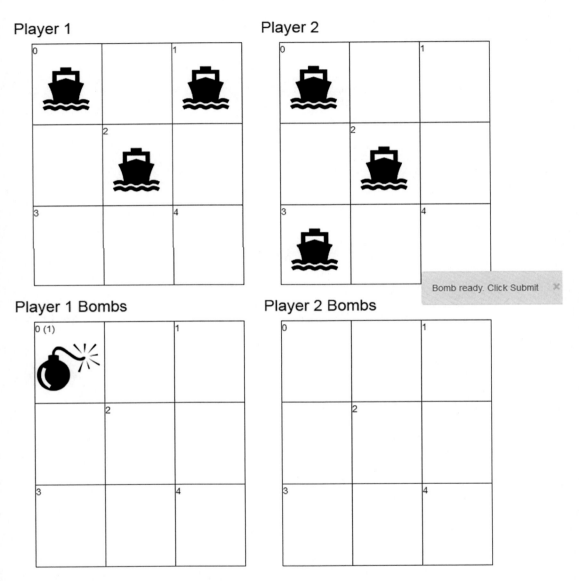

Figure 6-4. *Quantum Battleship user interface*

Listing 6-9. HTML Code for Figure 6-4

```html
<form id="frm1">
Device
<select id="device" name="device">
   <option value="local_qasm_simulator">Local Simulator</option>
   <option value="ibmqx_qasm_simulator">IBM Simulator</option>
   <option value="ibmqx2">ibmqx2</option>
</select>
   Place 3 ships per player, place a bomb & click submit.
<table>
    <tr>
    <td>
        <div><h3>Player 1</h3></div>
        <script type="text/javascript"> table(1, 's')</script>
    </td>
    <td>
        <div><h3>Player 2</h3></div>
        <script type="text/javascript"> table(2, 's')</script>
    </td>
    </tr>
    <tr>
    <td>
        <div><h3>Player 1 Bombs</h3></div>
        <script type="text/javascript"> table(1, 'b')</script>
    </td>
    <td>
        <div><h3>Player 2 Bombs</h3></div>
        <script type="text/javascript"> table(2, 'b')</script>
    </td>
    </tr>
</table>
</form>
```

The 3x3 boards are rendered dynamically using the document.write() system call as shown in Listing 6-10.

Listing 6-10. Dynamically Rendered Table Using document.write()

```
// type: 's' (ship) = checkbox, 'b' (bomb) = radio
function table (player, type) {
  var d     = document;
  var html  = '<table border="1">\n';
  var qubit = 0;

  for ( var i = 0 ; i < 3 ; i ++) {
    html += '<tr>';

    for ( var j = 0 ; j < 3 ; j ++) {
      if ( (i + j) % 2 == 0) {
        var id    = 'p' + player + type + qubit++;

        // checkbox = ship , radio = bomb
        var itype = type == 's' ? 'checkbox' : 'radio';
        var extra = type == 'b' ? ' onclick="cell_click_bomb(this)"'
              : ' onclick="return cell_click_ship(this)"';

        // <TD> SHIP-INDEX DAMAGE IMAGE </TD>
        html += '<td>' + (qubit - 1)
            + ' <span id="' + type + player + (qubit -1 ) + '"></span>'
            + '<input id="' + id + '" name="' + id + '" type="' + itype
            + '"' + extra + '>'
            + '<label for="' + id + '" class="ship"> </label></td>'
      }
      else {
        html += '<td> </td>';
      }
    }
    html += '</tr>\n';
  }
  html += '</table>';

  d.write(html);
}
```

Table 6-1 shows the major highlights of the user interface.

Table 6-1. *Cloud Battleship User Interface Tips and Tricks*

We hide the checkboxes and radio buttons using stylesheets. The selectors in lines 1 and 2 use the negation pseudo-class to hide the rule from older browsers. Lines 3 to 5 set the width, margin, and padding, in order to be able to position the alternative graphics accurately. Line 6 sets the opacity to render the standard user interface invisible.	```input[type=checkbox]:not(old),
input[type=radio   ]:not(old){
width   : 104px;
margin   : 0;
padding : 0;
opacity : 0;
}```
Each cell in the ships table displays • A qubit number • HTML span element to show damage percentage • An <INPUT type="checkbox"> modified to use a 100x100 pixel image instead of the usual control 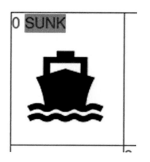	```input[type=checkbox]:not(old) + label
{
display   : inline-block;
margin-left : -104px;
padding-left : 104px;
background   : url('img/ship.png')
no-repeat 0 0;
line-height : 100px;
}``` We position the label and display the unchecked image. Line 2 sets the label to display as an inline-block element, allowing line 6 to set its height to the height of the alternative graphics and center the text vertically. Line 3 uses a negative margin to cover the space where the standard user interface would be displayed, and line 4 then uses padding to restore the label text to the correct position. The 104-pixel value used here is equal to the width of the image plus some additional padding so that the label text is not too close to the image. Line 5 displays the unchecked image in the space before the label text.

(continued)

Table 6-1. *(continued)*

Each bomb cell contains

- A qubit number
- HTML span element to show bomb counts for that position
- An <INPUT type="radio"> modified with CSS to use a 100x100 pixel image instead of the usual control

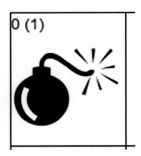

The style used to format the bomb is shown in the following fragment:

```
input[type=radio   ]:not(old) + label{
display      : inline-block;
margin-left  : -104px;
padding-left : 104x;
background   : url('img/bomb.png')
                no-repeat 0 0;
line-height  : 100px;
}
```

Next, we display the selected images when the checkboxes and radio buttons are selected:

```
input[type=checkbox]:not(old):checked
+ label{
background-position : 0 -100px;
}
input[type=radio]:not(old):checked +
label{
background-position : 0 -100px;
}
```

Because we have combined the images for the various states into a single image, the preceding rules modify the background position to show the appropriate image.

(continued)

Table 6-1. *(continued)*

The slick jQuery, Bootstrap, and Bootstrap-Growl libraries are used to render messages and debug information into the JS console:

```
<script type="text/javascript"
src="js/log.js"></script>
<script type="text/javascript"
src="js/jquery.js"></script>
<script type="text/javascript"
src="js/bootstrap.js"></script>
<script type="text/javascript"
src="js/bootstrap-growl.js"></
script>
<script type="text/javascript"
src="js/notify.js"></script>
```

The HTML is beautified using the quintessential Bootstrap library for GUI design. Messages are displayed on screen using the great JS library Bootstrap-Growl:

```
notify('Bomb ready. Click Submit',
info);
```

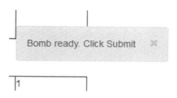

Game Rules and Validation

Because HTTP is a stateless protocol, all data structures and validation logic must be moved to the client. For example:

- Players cannot be allowed to place more than three ships on the board.

- No ship changes must be allowed after a bomb is placed.

- Bombs cannot be placed before all players place their ships.

- A global array of bomb counts is used to track user clicks: `var BOMBS = [[0,0,0,0,0], [0,0,0,0,0]]`. This array matches its Python counterpart: `bomb = [[0]*5 for _ in range(2)]`.

These rules can be enforced by adding a callback when a ship or bomb cell is clicked, respectively, as shown in Listing 6-11.

Listing 6-11. Enforcing Game Rules Using Click Callbacks from Book Source index.html

```
// Fires when a ship cell is clicked
function cell_click_ship ( obj ) {
    var id     = obj.id;
    var player = parseInt(id.charAt(1));
    var qubit  = parseInt(id.charAt(3));
    var json = countShipsBombs();

    LOGD('Cell Clicked ' + id  + ' Counts:' + JSON.stringify(json));
    if ( json.ships[0] > 3 || json.ships[1] > 3) {
        return error('All Players must place only 3 ships.');
    }
    // no ship changes after bombs are placed
    if ( json.bombs[0] > 0 || json.bombs[1] > 0 ) {
        return error('No ship changes after bombs are placed.');
    }
    return true;
}

// Fires when a bomb cell is clicked
function cell_click_bomb ( obj ) {
    var id     = obj.id;      // For Bombs: p[PLAYER]b[QUBIT]
    var player = parseInt(id.charAt(1));
    var qubit  = parseInt(id.charAt(3));

    // validate: { 'ships': [s1, s2], 'bombs': [b1, b2]}
    var json = countShipsBombs();
    LOGD('Bomb Clicked ' + id  + ' Counts:' + JSON.stringify(json));

    if ( json.ships[0] < 3 || json.ships[1] < 3) {
        $('#' + id).attr('checked', false);
        return error('All Players must place 3 ships first.');
    }
    if ( mustSubmit) {
        return error('Bomb in place already. Click Submit.');
    }
```

```
    // check player turn. Buggy?
    var dif = (json.bombs[player - 1] + 1) - json.bombs[ 1 - (player - 1)];

    if ( dif >= 2 ) {
      if ( BOMBS[player - 1 ][qubit] < 1 ) {
       $('#' + id).attr('checked', false);
      }
      return error("Not your turn. It's player " + ((1-(player-1)) + 1) );
    }

    // Count bomb
    BOMBS[player - 1 ][qubit]++;

    // Assign counts to: d[PLAYER][QUBIT]
    $('#b' + player + qubit).html("(" + BOMBS[player - 1 ][qubit] + ")");

    // bomb in place, click submit
    notify('Bomb ready. Click Submit', 'info');
    mustSubmit = true;
}

function error (msg) {
    notify(msg, 'danger');
    return false
}
```

Now the data can be sent to the backend for consumption.

End Point and Response Format

Each request must be sent to the Web Server asynchronously using AJAX. Furthermore a specific format must be used for the query string. Thus the request-response format is as follows:

Given the end point `http://localhost/~centos/battleship/cgi-bin/qiskit-driver.sh`, we assume that

- The username is centos.

- The code has been deployed into the user's home folder: $HOME/centos/public_html/battleship.

- Python 3 must be activated by using the wrapper script qiskit-driver. sh. This is required only if multiple versions of Python are present in the host (see "Running Multiple Versions of Python" section).

The following values are required in the request query string:

- *ships1*: comma-separated list of three positions for player 1 ships
- *ships2*: comma-separated list of three positions for player 2 ships
- *bombs1*: comma-separated list of five bomb counts for player 1
- *bombs2*: comma-separated list of five bomb counts for player 2
- *device*: quantum device. For example, local_qasm_simulator, ibmq_ qasm_simulator, or ibmqx2.

For example, here is a full AJAX request to be run in the IBM remote simulator:

```
http://localhost/~centos/battleship/cgi-bin/qiskit-driver.sh?ships1=0,1,2&s
hips2=0,1,2&bombs1=0,1,0,0,0&bombs2=0,0,0,0,0&device=ibmqx_qasm_simulator
```

When the player clicks Submit, the ship positions, ships1 and ships2, and bomb counts, bombs1 and bombs2, are assembled from the DOM tree and the global BOMBS array. The request end point and query string are defined, and the HTTP GET request is sent via AJAX as shown in Listing 6-12.

Listing 6-12. Submitting Data to the Backend from Book Source index.html

```
function submit() {
  var frm  = $('#frm1');
  var url  = "cgi-bin/qiskit-driver.sh";

  // Data format: ships1=0,1,2&ships2=0,1,2&bombs1=0,1,0,0,0&bom
  bs2=0,0,0,0,0
  // ships has the positions per player, bombs has the bomb position counts
  per player
  // ships: 3 ships per player, bombs: 5 position counts
  var data = ";
  var s1   = ";
  var s2   = ";
```

```
for ( var i = 0 ; i < 5 ; i++) {
  if ( $('#p1s' + i).prop('checked') ) s1 += ',' + i;
  if ( $('#p2s' + i).prop('checked') ) s2 += ',' + i;
}
// remove 1st comma
if (s1.length > 0) s1 = s1.substring(1);
if (s2.length > 0) s2 = s2.substring(1);

// query string
data = 'ships1=' + s1 + '&ships2=' + s2
    + '&bombs1=' + BOMBS[0].join(',') + '&bombs2=' + BOMBS[1].join(',')
    + '&device=' + $('#device').val();

LOGD('Url:' + url + ' data=' + data);

// https://api.jquery.com/jquery.get/
$.get( url, data)
.done(function (json) {
  handleResponse (json);
})
.fail(function() {
    LOGD( "error" );
    notify('Internal Server Error. Check logs.', 'danger');
})
}
```

If something goes wrong, an error notification will be displayed on screen, else the expected response JSON will be sent to a handler for consumption. Let's see how.

Response Handler

The job of the response handler is to consume the backend response and update damage counts, display error messages if any, or repeat this process until a winner is declared. Listing 6-13 shows this process, but before we do, let's take a look at the critical format of the response JSON:

```
{"status":200,"message":"OK","damage":[[0.475,0,0,0.70,0],[0.786,0.90,0,0,0.]]}
```

The most important key is *damage*. It contains a 2D array representing ship damage positions for each player. The damage is a percentage between 0 and 1. This data is used by the response handler to update the user interface.

Listing 6-13. Response Handler from Book Source index.html

```
function handleResponse (json) {
  LOGD("Got: " + JSON.stringify(json))
  var damage   = json.damage;
  var d1       = damage[0];  // damage P1
  var d2       = damage[1];  // damage P2

  for ( var i = 0 ; i < 5 ; i++) {
      var pct1 = (d1[i] * 100).toFixed(1);
      var pct2 = (d2[i] * 100).toFixed(1);
      var s1, c1, s2, c2;
      if ( pct1 < 90 ) {
          s1 = '['  + pct1 + '%]';
          c1 = 'cyan';
      }
      else {
          s1 = 'SUNK';
          c1 = 'red';
          notify('Player 1 Ship ' + i + ' sunk.', 'warning');
      }
      if ( pct2 < 90 ) {
          s2 = '['  + pct2 + '%]';
          c2 = 'cyan';
      }
      else {
          s2 = 'SUNK';
          c2 = 'red';
          notify('Player 2 Ship ' + i + ' sunk.', 'warning');
      }
          //LOGD(i + ' s1=' + s1 + ' s2=' + s2 + ' d1=' + d1[i] +
          ' d2=' + d2[i]);
```

```
    $('#s1' + i).html(s1).css('background-color', c1);
    $('#s2' + i).html(s2).css('background-color', c2);
}

// Game Result: damage sum > 2.85 (0.95 * 3) = loss
// https://www.w3schools.com/jsref/jsref_reduce.asp
// array.reduce(function(total, currentValue, currentIndex, arr),
initialValue)
var s1 = d1.reduce(function(total, currentValue, currentIndex, arr)
    { return total + currentValue}, 0);
var s2 = d2.reduce(function(total, currentValue, currentIndex, arr)
    { return total + currentValue}, 0);
var winner = 0;
if ( s1 > 2.85) winner = 2;
if ( s2 > 2.85) winner = 1;

LOGD ("Results Damage sums s1:" + s1 + " s2:" + s2);
if ( winner != 0) {
    notify ('** G.A.M.E O.V.E.R Player ' + winner + ' wins **',
    'success');
    gameover = true;
}

// enable submit
$("#btnSubmit").prop("disabled", false);
}
```

- The handler extracts the damage array and loops for each position converting the damage for each player to a 1-100 percentage.

- Colorization is used to display the damage percentage for a dramatic effect. Messages are displayed on screen for each ship sunk (see Figure 6-5).

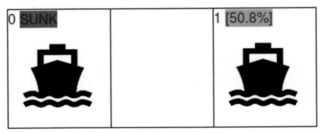

Figure 6-5. *Damage colorization*

- If the damage sum exceeds a 90% threshold for all ships of a player, a winner is declared and the game is over. Click the *Reset* button to start a new game.

To reset the game, we just uncheck all checkboxes and radio buttons and reset the global BOMBS array as shown in Listing 6-14.

Listing 6-14. Game Reset from Book Source index.html

```
// Restart game: fires when the reset button is clicked
function reset_click () {
  if ( ! confirm("Are you sure?")) {
    return;
  }
  gameover = false;
  for ( var i = 0 ; i < 5 ; i++) {
    $('#p1s' + i).attr('checked', false);
    $('#p2s' + i).attr('checked', false);
    $('#p1b' + i).attr('checked', false);
    $('#p2b' + i).attr('checked', false);
    // info spans
    $('#s1' + i).html(");
    $('#s2' + i).html(");
    $('#b1' + i).html(");
    $('#b2' + i).html(");
```

```
   BOMBS[0][i] = 0;
   BOMBS[1][i] = 0;
  }
}
```

Now it is time to run, deploy, test, and troubleshoot if necessary. I use CentOS 6 for development which bundles Python 2.7 as default. Remember that we must run in Python 3.5 or later.

Tip   Listings 6-9 thru 6-12 can be found at the book source under `Workspace\ Ch06\battleship\index.html` as well as all resources required to deploy the game on the cloud.

Running Multiple Versions of Python

Chapter 3 explains how to install and run both Python 3.6 and Python 2.7 separately. In this particular case, a wrapper script is used to activate Python 3 in the CGI backend before invoking the quantum program.

```
#!/bin/sh
# home dir
root=/home/centos
program=qbattleship.py

# Activate python 3
source $root/qiskit/qiskit/bin/activate

# execute python quantum program
python $program
```

The previous script simply activates Python 3 and invokes the real quantum program qbattleship.py. It is required, or else the Web Server will use the default Python installation (2.7) and the program will fail as the QISKit requires Python 3.5 or later. Remember that a Python 3 environment was created in the user's home as follows:

```
$ mkdir -p $HOME/qiskit
$ cd $HOME/qiskit
$ python3.6 -m venv qiskit
```

To activate the virtual environment:

```
$ source qiskit/bin/activate
```

Now finally deploy and test. Hopefully troubleshooting won't be necessary.

Solution 3: Deploy and Test

In this section we deploy the game in the Apache HTTPD server and look at the game in action. The full source for the game including all support files, styles, images, CGI wrapper, and quantum program, can be found in the book source under `Workspace\Ch06\battleship`. The folder layout is shown in Figure 6-6.

Note   This section assumes that you already have installed Apache HTTPD in your system and that the default service has been configured and it is running properly. If this is not the case, there are plenty of tutorials up there. For example, for CentOS 7 I like `www.liquidweb.com/kb/how-to-install-apache-on-centos-7/`.

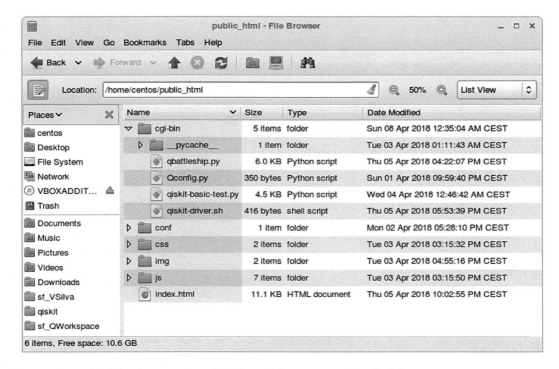

Figure 6-6. *Folder layout for the Cloud Quantum Battleship*

1. Create a folder called public_html in your user's home.

   ```
   $ mkdir $HOME/public_html
   ```

2. Create the cgi-bin folder under public_html to contain the CGI Python scripts.

   ```
   $ mkdir $HOME/public_html/cgi-bin
   ```

3. Configure the HTTPD server to enable access from the user's public_html as well as public_html/cgi-bin folders (see Listing 6-15). Note that cgi-bin needs a special permission to allow for CGI script execution.

4. If you wish to use the book source, copy all files from Workspace\ Ch06\battleship to public_html/battleship.

5. Make sure the file permissions are correct for the public_html folder and all subfolders and files. This is very important; if the permissions are incorrect, the browser will give a "500 – Internal Server Error" response. This was a major source of headaches when I was testing in my CentOS 6 desktop: `$ chmod -R 755 public_html`.

Listing 6-15. Configuration to Enable HTTP Requests from the User's public_html Directory (CentOS 6/Apache HTTPD 2.2)

```
<IfModule mod_userdir.c>
    #UserDir disabled
    #
    # To enable requests to /~user/ to serve the user's public_html
    # directory, remove the "UserDir disabled" line above, and uncomment
    # the following line instead:
    #
    UserDir public_html
</IfModule>

<Directory /home/*/public_html>
    AllowOverride FileInfo AuthConfig Limit
    Options MultiViews Indexes SymLinksIfOwnerMatch IncludesNoExec +ExecCGI
```

```
    AddHandler cgi-script .cgi
    <Limit GET POST OPTIONS>
        Order allow,deny
        Allow from all
    </Limit>
</Directory>

<Directory "/home/*/public_html/cgi-bin">
    AllowOverride None
    Options ExecCGI
    SetHandler cgi-script
</Directory>
```

Tip   Enabling requests from *public_html* (Listing 6-15) requires the Apache module *userdir* to be enabled in *httpd.conf* (uncomment `LoadModule userdir_module modules/mod_userdir.so`). This module may not be enabled by default.

Copy the script in Listing 6-15 to the system folder `/etc/httpd/conf.d`. This folder contains configuration files loaded automatically by Apache HTTPD at startup. Now start the HTTPD server, in CentOS (note that this assumes that you already have Apache HTTPD installed in your system):

```
$ sudo service httpd start        (CentOS 6)
$ sudo systemctl start httpd      (CentOS 7)
```

Finally, for the grand finale, start your browser and open the URL `http://localhost/~centos/battleship/` (assuming the username is *centos*). Hopefully there will be no issues and you can start playing Quantum Battleship in the cloud; however if something goes wrong, here is a list of issues that I came across setting things up.

Troubleshooting

Most of the issues I faced were related to file permissions due to my rustiness using the good old Apache HTTPD:

- *Apache HTTPD idiosyncrasies*: Enabling requests from the user's home (Listing 6-15) requires the module *userdir* to be enabled in the daemon configuration *httpd.conf*. Depending on your OS, this module may not be enabled by default. Also HTTPD 2.4 users: Listing 6-15 is for Apache v2.2; v2.4 may require a different syntax.

- *HTTP status 500 - Internal Server Error in the Browser*: Make sure the file permissions for public_html and all files and subfolders are set to 755. You can diagnose this by looking at the HTTPD log files located at

    /var/log/httpd/error_log
    /var/log/httpd/suexec.log

For example, here is a snippet from suexec.log telling me my permissions were messed up:

```
$ tail -f /var/log/httpd/suexec.log
[2018-04-02 17:03:45]: cannot get docroot information (/home/centos)
[2018-04-02 17:10:13]: uid: (500/centos) gid: (500/centos) cmd: first.cgi
[2018-04-02 17:10:13]: directory is writable by others: (/home/centos/
public_html)
```

Tip   Apache suEXEC is a feature of the Apache Web Server. It allows users to run CGI and SSI applications as a different user. In CentOS suEXEC writes a log to /var/log/httpd/suexec.log.

- *SELinux headaches*: This is a Linux kernel security module that provides a mechanism for supporting access control security policies. In CentOS this feature is enabled by default. It can be disabled temporarily from the command line using the command:

    ```
    $ sudo setenforce 0
    ```

or permanently by editing the file /etc/sysconfig/selinux and setting the value of the
SELINUX key to *disabled*.

```
$ sudo vi /etc/sysconfig/selinux
SELINUX=disabled
SELINUXTYPE=targeted
```

Note that SELinux can cause trouble when invoking the CGI scripts or when the
quantum program attempts to execute remote code against the IBM simulator or real
device.

- *Python bugs*: If any error occurs in the Python script, the CGI exception
  handler will catch it and dump a nice HTML page in the current
  working directory (cgi-bin). Figure 6-7 shows the output from a timeout
  error occurred when executing in the real quantum device *ibmqx2*.

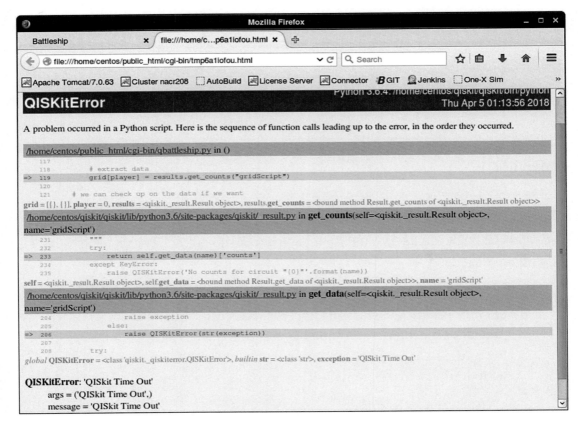

Figure 6-7. *Python error dump created by the cgi package*

- *API configuration issues*: Finally if running in a real quantum device, make sure the configuration is correct in Qconfig.py (including the API token for your Q Experience account) as shown in the next fragment:

```
APItoken = 'YOUR-API-TOKEN'
config = {
    'url': 'https://quantumexperience.ng.bluemix.net/api',
}
```

Note that Qconfig.py must live in the same location as the quantum program qbattleship.py, that is, the *cgi-bin* folder. Still, further improvements can be made to the game; let's discuss them in the next section.

Further Improvements

Cloud Battleship from the previous section can use some improvements which you could probably notice after playing the game for a while. Here is a list of my ideas:

- The user interface shows the ship and bomb boards for both players. In a real battleship game each player should open its own browser window, set his ships, and start bombing the opponent.

- *Gate state*: ship, bomb positions, and quantum device are all kept in the client due to the fact that HTTP is a stateless protocol. That is, a request comes, the python program runs, and a response is sent back. After that, all memory disappears. A real game should use a server-based player lobby to host all the game state (e.g., using an application server) and coordinate communication between browser windows.

A Better Cloud Battleship

The ultimate Cloud Quantum Battleship game should use two browser windows for each player with ship and bomb boards each. Furthermore the Apache HTTPD should be replaced with an application server (such as Apache Tomcat) which is capable of storing game state in the server. The layout for this is shown in Figure 6-8.

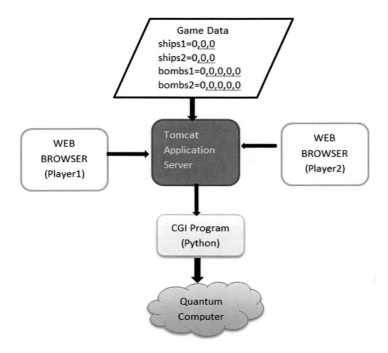

Figure 6-8. *Improved Cloud Quantum Battleship*

- A rudimentary game lobby can be implemented as a Tomcat web application to store ship, bomb positions, and quantum device.

- The web application can use the host OS runtime facilities (in this case Java's runtime API) to invoke the quantum Python script, get damage results back, and dispatch them to each player.

- To avoid the always-annoying browser page refreshes, each browser can connect via WebSocket to the application server. This will keep a permanent connection open where JSON text messages can be sent quickly between clients.

Connecting via WebSocket

The UI web page needs to be modified slightly to connect via WebSocket instead of AJAX as shown in Listing 6-16.

Tip   An Eclipse project for this section is provided in the book source under
`Workspace\Ch06_BattleShip`. Given all the complex parts of this web app,
I recommend that you open the workspace in your IDE and read along. Note that
I assume that you have a good level of proficiency on writing apps with the
Eclipse/Tomcat combo.

Listing 6-16. WebSocket Javascript Client Code Under WebContent/js/
websocket.js

```
// Server WS endpoint (file: websocket.js)
var END_POINT = "ws://localhost:8080/BattleShip/WSBattleship";

// Random ID used to track a client
var CLIENT_ID = Math.floor(Math.random() * 10000);

function WS_connect(host) {
  LOGD("WS Connect " + host);

  if ('WebSocket' in window) {
    this.socket = new WebSocket(host);
  } else if ('MozWebSocket' in window) {
    this.socket = new MozWebSocket(host);
  } else {
    LOGE('Error: WebSocket is not supported by this browser.');
    return;
  }

  this.socket.onopen = function() {
    LOGD('WS Opened ' + host);
  };

  this.socket.onclose = function() {
    LOGD('WS Closed ' + host);
  };

  this.socket.onmessage = function(message) {
    // { status: 200 , message :'...'}
```

```
    LOGD('OnMessage: ' + message.data);
    var json = JSON.parse(message.data);

    if ( json.status >= 300 && json.status < 400) {
      // warning
      notify(json.message, 'warning');
    }
    if ( json.status >= 400 ) {
      // error
      notify(json.message, 'danger');
      return;
    }
    handleResponse (json);
  };
}
function WS_initialize () {
  var clientId = CLIENT_ID;
  var host     = END_POINT;
  this.url     = host + '?clientId=' + clientId;

  WS_connect(this.url);
};
function WS_send (text) {
  this.socket.send(text);
};
```

In the client:

- All major browsers implement the WebSocket standard used to keep a persistent connection against a capable server. For this, an end point URL of the form ws://localhost:8080/BattleShip/WSBattleship is created in line 2. Note that parameters can be sent in WebSocket end points just like regular URLs. Thus the final WS URL is ws://localhost:8080/BattleShip/WSBattleship?clientId=RANDOM-ID where a random ID is used to track each player.

- WebSocket in Javascript uses a callback system to receive events such as

  - *socket.onopen*: It fires when the socket is opened. Line 23 shows the callback used to handle this event.

  - *socket.onclose*: It fires when the connection is broken: when the browser is closed or refreshed or the server dies for example.

  - *socket.onmessage*: This is the most important callback. It fires when a message is received, and it is used to consume the JSON message sent by Python, just as AJAX does in the previous version.

When the player browser page loads, the client connects using the DOM `window.onload` callback:

```
function win_onload () {
    WS_initialize ();
}
window.onload = win_onload;
```

In the server we need a WebSocket-capable application server. Luckily for us, Tomcat fully implements the WebSocket standard in all operating systems. Listing 6-17 shows a basic implementation of a WebSocket handler in Java.

Listing 6-17. WebSocket Server Handler Skeleton (WSConnector.java)

```
@ServerEndpoint(value = "/WSBattleship")
public class WSConnector {
  // connections
  private static final List<WSConnectionDescriptor> connections =
    new CopyOnWriteArrayList<WSConnectionDescriptor>();

  // Game data Player-ID => {name: 'Player-1', ships: "0,0,0:, bombs:
  "0,0,0,0,0 }
  private static final Map<String, JSONObject> data =
   new HashMap<String, JSONObject>();

  /** The client id for this WS */
  String clientId;
```

```java
private String getSessionParameter (Session session, String key) {
  if ( ! session.getRequestParameterMap().containsKey(key)) {
    return null;
  }
  return session.getRequestParameterMap().get(key).get(0);
}

@OnOpen
public void open(Session session) {
  clientId = getSessionParameter(session, "clientId");

  // no duplicates?
  WSConnectionDescriptor conn = findConnection(clientId);

  if ( conn != null) {
    unicast(conn.session,
      WSMessages.createStatusMessage(400
          , "Rejected duplicate session.").toString());
  }
  else {
    connections.add(new WSConnectionDescriptor(clientId, session));
  }
  dumpConnections("ONOPEN " +  clientId );
}

@OnClose
public void end() {
}

@OnMessage
public void incoming(String message) {
  WSConnectionDescriptor d = findConnection(clientId);
  try {
      JSONObject root = new JSONObject(message);
      String name    = root.getString("name");
      String action  = root.optString("action");
```

```
    .  // reset game?
       if ( action.equalsIgnoreCase("reset")) {
          multicat(WSMessages.createStatusMessage(300
             , "Game reset by " + name).toString());
          data.clear();
          return;
       }

       // Validate game rules...
       // Execute python script
       linuxExecPython(args);

  } catch (Exception e) {
    LOGE("OnMessage", e);
  }
}

@OnError
public void onError(Throwable t) throws Throwable {
  LOGE("WSError: " + t.toString());
}
}
```

In Java, WebSocket server handlers are implemented using the J2EE annotation standard. This makes code reusable across all vendors.

- In Listing 6-17 line 1, the Java class WSConnector defines the annotation @ServerEndpoint(value = "/WSBattleship"). This powerful instruction is all we need to build a server handler. The value WSBattleship is the name of the handler; thus the full server end point will be ws://host:POT/Battleship/WSvattleship?QUERY-STRING.

- Callbacks for open, close, and message events are declared using the annotations: @OnOpen, @OnClose, and @OnMessage, respectively. Note that the method names are irrelevant, what matters are the parameters:

    - *OnOpen*: Receives a Session object which contains information about the connection.

    - *OnClose*: No parameters in this one. It fires when the browser connection dies.

    - *OnMessage*: The most important of the lot. It fires when a text message is sent by a client with the data as the argument.

- Keep in mind that a single instance of the WSConnector class will be created for each client connection; thus in line 5, we use a thread safe static list List<WSConnectionDescriptor> connections to track all client connections. Line 8 declares a static hash map to track game data with the key being a player id and the value, a JSON object sent by the browser. For example, [Player-1 => {name: 'Player-1', ships: "0,0,0:, bombs: "0,0,0,0,0", device: "local_qasm_simulator"}].

- When the message callback fires (file WSConnector.java lines 201-253), the text message is parsed as JSON, the data is stored in memory, game rules applied, and if everything is correct, the Python script is invoked with ship and bomb positions. Finally the results are collected and sent back to each client for update.

To send a message back to the client, the Session object can be used:

```
session.getBasicRemote().sendText("Some Text")
```

To send a message to everybody (multicast), the connections list can be used:

```
static void multicast ( String message ) {
  for ( WSConnectionDescriptor conn : connections) {
    conn.session.getBasicRemote().sendText(message)
  }
}
```

Invoking Python and Setting File Permissions from Java

Even though the Java language is designed to be OS agnostic, invoking operating system commands is possible via the `Runtime.getRuntime().exec("command")` system call. Listing 6-18 shows a very simple class to execute a command and read its standard output into a string buffer.

Listing 6-18. Executing OS Commands and Extracting Results (SysRunner.java)

```java
public class SysRunner {
  final String command;
  final StringBuffer stdout = new StringBuffer();
  final StringBuffer stderr = new StringBuffer();

  public SysRunner(String command) {
    this.command = command;
  }

  public void run () throws IOException, InterruptedException {
    final Process process   = Runtime.getRuntime().exec(command);
    pipeStream(process.getInputStream(), stdout);
    pipeStream(process.getErrorStream(), stderr);
    process.waitFor();
  }

  private void pipeStream (InputStream is, StringBuffer buf) throws
  IOException {
    BufferedReader br = new BufferedReader(new InputStreamReader(is));
    String line;

    while ((line = br.readLine()) != null) {
      buf.append(line);
    }
  }
  public StringBuffer getStdOut () {
    return stdout;
  }
}
```

```
public StringBuffer getStdErr () {
  return stderr;
}
}
```

To get the output from a command, use the process input stream, read from it, and store the data in a string buffer (lines 17-24 of Listing 6-18): `pipeStream(process.getInputStream(), stdout)`. Now we have the tool to execute the Python program but still need to deal with the Linux file permissions. Remember that the Python script must be included in the web application itself (see Figure 6-9). Therefore when the application server extracts the Battleship web app in the file system along with the Python code, the script will take the default file permission of 644 (not world executable). This will cause the script to fail when run.

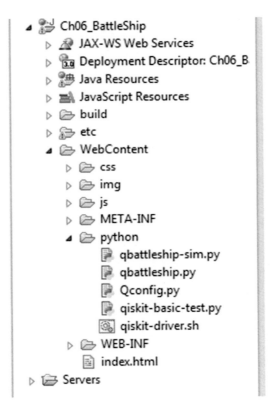

Figure 6-9. Project layout of the Cloud Battleship J2EE project

To fix the file permissions for the Python code within a web app, execute a chmod OS command with the file names as shown in the next paragraph:

```
// Get the base path for the python code
// ...webapps/BattleShip/python/
String root = IOTools.getResourceAbsolutePath("/") + "../../";

// Special *&$# chars don't work
String cmd = "/bin/chmod 755 " + base +  "python" + File.separator;

String[] names = { "Qconfig.py", "qiskit-basic-test.py"
  , "qiskit-driver.sh", "qbattleship-sim.py", "qbattleship.py"};

for (int i = 0; i < names.length; i++) {
  SysRunner r = new SysRunner(cmd + names[i]);
  r.run();
}
```

The base path of the app installation can be obtained in Java using reflection as shown in the following:

```
public static String getResourceAbsolutePath(String resourceName) throws
UnsupportedEncodingException {
    URL url    = IOTools.class.getResource(resourceName);
    String path = URLDecoder.decode(url.getFile(), DEFAULT_ENCODING);

    // path -> Windows: /C:/.../Workspaces/.../
    // path-> Linux:   /home/users/foo...
    if ( path.startsWith("/") && OS_IS_WINDOWS) {
      // gotta remove the first / in Windows only!
      path = path.replaceFirst("/", "");
    }
    return path;
}
```

Finally the Python quantum program can be executed from the WebSocket message callback as shown in Listing 6-19.

Listing 6-19. Executing the Quantum Program and Sending Results Back

```java
// Args: ships1=0,0,0 ships2=0,0,0 bombs1=0,0,0,0,0 bombs2=0,0,0,0,0
// device=local_qasm_simulator
private void linuxExecPython (String args) throws Exception {
    // STDOUT {status: 200, message: 'Some text', damage:
    [[0,0,0,0,0],[0,0,0,0,0]]}
    StringBuffer stdout = IOTools.executePython(SCRIPT_ROOT, args);
    JSONObject resp = new JSONObject(stdout.toString());

    // send back to clients in reverse order
    JSONArray damage = resp.getJSONArray("damage");
    resp.remove("damage");
    final int size = damage.length() - 1;

    for (int i = 0; i < connections.size(); i++) {
        resp.put("damage", damage.get( size - i));
        unicast(connections.get(i).session, resp.toString());
        resp.remove("damage");
    }
}
// base: WEPAPP_PATH/python/qiskit-driver.sh
// args: WEPAPP_PATH/python/qbattleship.py
//         0,0,0 0,0,0 0,0,0,0,0 0,0,0,0,0 device
public static StringBuffer executePython (String base, String args)
throws IOException, InterruptedException {
    String driver = base + File.separator + "python" + File.separator +
    "qiskit-driver.sh";
    String program = base + File.separator + "python" + File.separator +
    "qbattleship.py";
    String cmd = driver + " " + program + ( args != null ? " " + args : "");

    SysRunner r = new SysRunner(cmd);
    r.run();
    return r.getStdOut();
}
```

To execute the Python quantum program, the code in Listing 6-19

- Obtains the LOCATION of the python folder within the web app. That is `TOMCAT-ROOT/webapps/Battleship/python`

- Executes the driver script `LOCATION/qiskit-driver.sh LOCATION/qbattleship.py` with the arguments:

  - *ships1*: Ship locations for player 1

  - *ships2*: Ship locations for player 2

  - *bombs1*: Bomb counts for player 1

  - *bombs2*: Bomb counts for player 2

  - *device*: Quantum device

- Dispatches the results back to the clients

Finally, from your IDE, export the Cloud Quantum Battleship web archive (WAR), deploy it in your Tomcat container, and game on with two web browsers at `http://localhost:8080/BattleShip/` (see Figure 6-10). I have assumed that you are proficient in doing this but just in case:

1. Export the web app as a web archive WAR, and right-click the Ch06_Battleship project in your IDE (see Figure 6-9). Click *Export > Web Archive*, and select a name/destination (e.g., Ch06_Battleship.war).

2. Make sure the Tomcat service is up and running. If not installed by default in your system, here is some help:

```
yum -y install java (CentOS 6,7)
yum -y install tomcat7 tomcat7-webapps tomcat7-admin-webapps
(CentOS 6,7)

service tomcat7 start (CentOS 6)
systemctl start tomcat7 (CentOS 7)
```

3. Use the Tomcat manager UI at `http://yourhost:8080/manager/` to upload and deploy the archive into your Linux Tomcat container. (Tip: The manager will ask for a user/password; if you don't have them, edit the file /etc/tomcat7/tomcat-users.xml).

4. You should now be able to point two browsers to `http://localhost:8080/BattleShip/` (Tip: Tomcat web apps are deployed into the folder /var/lib/tomcat7/webapps). Having trouble? Check the container logs at /var/log/tomcat7/catalina.out.

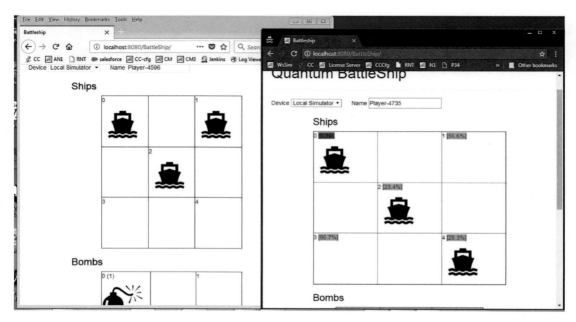

Figure 6-10. *Improved Cloud Battleship with two browsers*

This chapter has shown how the popular game Battleship can be run in a quantum computer using a single-qubit partial NOT gate to compute ship damage. For this purpose, the Quantum Battleship sample from the QISKit tutorial has been used. Furthermore, the game has been taken to the next level by giving it a major face lift. You have learned how this quantum code can be invoked in the cloud using CGI scripting via the Apache HTTPD server. Further improvements have been made to play using two browsers via the Tomcat J2EE container. The code for both projects is available as Eclipse projects from the book source at `Workspace\Ch06` and `Workspace\Ch06_BattleShip`, respectively.

CHAPTER 6   FUN WITH QUANTUM GAMES

The next chapter explores two game puzzles that show the remarkable power of quantum algorithms over their classical counterparts: the counterfeit coin puzzle and the Mermin-Peres Magic Square. These are examples of quantum pseudo-telepathy or the ability of players to achieve outcomes only possible if they were reading each other's minds during the game.

## CHAPTER 7

Game Theory: With Quantum Mechanics, Odds Are Always in Your Favor

This chapter explores two game puzzles that show the remarkable power of quantum algorithms over their classical counterparts:

- *The counterfeit coin puzzle*: It is a classical balance puzzle proposed by mathematician E. D. Schell in 1945. It is about balancing coins to determine which holds a different value (counterfeit) using a balance scale and a limited number of tries.

- *The Mermin-Peres Magic Square game*: This is an example of quantum pseudo-telepathy or the ability of players to achieve outcomes that would only be possible if they mysteriously communicate during the game.

In both cases, quantum computation gives quasi-magical abilities to the players, just as if they were cheating all along. Let's see how.

© Vladimir Silva 2018
V. Silva, *Practical Quantum Computing for Developers*, https://doi.org/10.1007/978-1-4842-4218-6_7

Counterfeit Coin Puzzle

In this puzzle, the player has eight coins and a beam balance. One of the coins is fake and thus underweight. The goal of the game is to figure out which coin is fake by using the balance only twice. Can you figure out how? Let's run through the solution shown in Figure 7-1.

1. Given eight coins, put coins 1-3 on the left side of the balance and 4-6 on the right side. Leave the last two coins 7 and 8 on the side and weight.

2. If the balance leans right, the counterfeit is among 1-3 (left). Remember that the fake coin is lighter. Thus take out the last coin from the left (3) and weight again (for the second time).

   • If the beam leans right, the counterfeit is 1. Stop.

   • If the beam leans left, the counterfeit is 2. Stop.

   • If the beam is balanced, the counterfeit is 3. Stop.

3. If the balance leans left, the counterfeit is among 4-6. Take out the last coin (6) and weight again.

   • If the beam leans right, the counterfeit is 4. Stop.

   • If the beam leans left, the counterfeit is 5. Stop.

   • If the beam is balanced, the counterfeit is 6. Stop.

4. If the beam is balanced, the counterfeit is either 7 or 8. Put 7 and 8 in the balance and weight again.

   • If the beam leans right, the counterfeit is 7. Stop.

   • If the beam leans left, the counterfeit is 8. Stop.

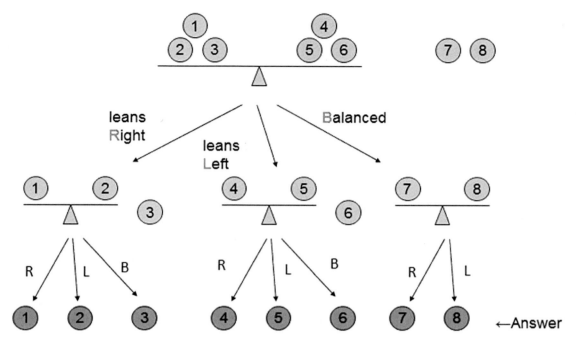

Figure 7-1. Counterfeit puzzle solution for eight coins

From the procedure in the previous section, a classical algorithm can be implemented independent of the total number of coins N and the number of counterfeit coins k. In general terms, the time complexity of the generalized counterfeit coin puzzle is given by

$$O\big(k\log(N/k)\big)$$

Tip   It has been proven that the minimal number of tries required to find a single counterfeit coin using the balance beam in a classical computer is two.

Counterfeit Coin, the Quantum Way

Believe it or not, there is a quantum algorithm that can find the counterfeit using a quantum balance only once, independent of the number of coins N! In general terms, for any number of counterfeit coins k, independent of N, the time complexity of such algorithm is given by

$$O\left(k^{1/4}\right)$$

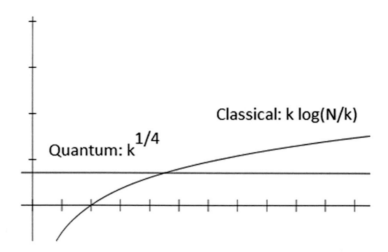

Figure 7-2. *Quantum vs. classical time complexities for the counterfeit coin puzzle*

Tip   The quantum counterfeit coin algorithm is an example of quartic speedup over its classical counterpart.

Thus Figure 7-2 shows the power of a quantum algorithm over its classical counterpart for the counterfeit coin puzzle. Now, let's dig deeper. A quantum algorithm to find a single counterfeit coin (k = 1) can be summarized in three stages: query the quantum beam balance, construct the quantum balance, and identify the false coin.

Step 1: Query the Quantum Beam Balance

A quantum algorithm will query the beam balance in superposition. To do this, we use a binary query string to encode coins placed on the pans. For example, the query string 11101111 means all coins are on the beam except coin with index 3. The beam is balanced when no false coin is included and tilted otherwise. The next table illustrates this.

N (# of coins)	F (index of false coin)	Query string	Result
8	3	11101111	Balanced (0)
8	3	11111111	Tilted (1)

The procedure can be summarized as follows:

1. Use two quantum registers to query the quantum balance, where the first register is for the query string and the second register for the result.

2. Prepare the superposition of all binary query strings with even number of 1s.

3. To obtain the superposition of states of even number of 1s, perform a Hadamard transform on the basis state |0>, and check if the Hamming weight of |x| is even. It can be shown that the Hamming weight of |x| is even if and only if $x1 \oplus x2 \oplus \ldots \oplus xN = 0$.

Tip   The Hamming weight (hw) of a string is the number of symbols that are different from the zero symbol of the alphabet used. For example, hw(11101) = 4, hw(11101000) = 4, hw(000000) = 0.

---

4. Finally, measure the second register, and if |0> is observed, then the first register is the superposition of all binary query strings we want. If we get |1> then repeat the procedure until |0> is observed.

Note that each repetition is guaranteed to succeed with probability exactly half. Hence, after several repetitions, we should be able to obtain the desired superposition state. Listing 7-1 shows an implementation of a quantum program to query the beam balance with the corresponding graphical circuit shown in Figure 7-3.

Note   For the sake of clarity, the full counterfeit coin program has been broken in Listings 7-1 thru 7-3. Although you should be able to join the sections to run the program, a full listing is available from the source at `Workspace\Ch07\p_counterfeitcoin.py`.

Listing 7-1. Script to Query the Quantum Beam Balance

```
# ------- Query the quantum beam balance
Q_program = QuantumProgram()
Q_program.set_api(Qconfig.APItoken, Qconfig.config["url"])

# Create numberOfCoins +1 quantum/classic registers
# 1 extra qubit for recording the result of quantum balance
qr = Q_program.create_quantum_register("qr", numberOfCoins +1 )

# for recording the measurement on qr
cr = Q_program.create_classical_register("cr", numberOfCoins + 1)

circuitName = "QueryStateCircuit"
circuit   = Q_program.create_circuit(circuitName, [qr], [cr])

N = numberOfCoins

#Create uniform superposition of all strings of length N
for i in range(N):
  circuit.h(qr[i])

#Perform XOR(x) by applying CNOT gates sequentially from qr[0] to qr[N-1]
# and storing the result to qr[N]
for i in range(N):
  circuit.cx(qr[i], qr[N])
```

```
# Measure qr[N] and store the result to cr[N].
# continue if cr[N] is zero, or repeat otherwise
circuit.measure(qr[N], cr[N])

# query the quantum beam balance if the value of cr[0]...cr[N] is all 0
# by preparing the Hadamard state of |1>, i.e., |0> - |1> at qr[N]
circuit.x(qr[N]).c_if(cr, 0)
circuit.h(qr[N]).c_if(cr, 0)

# rewind the computation when cr[N] is not zero
for i in range(N):
  circuit.h(qr[i]).c_if(cr, 2**N)
```

Figure 7-3 shows a complete circuit for counterfeit coin puzzle for eight coins, one counterfeit at index 6. The circuit displays all the stages described here for the IBM Q Experience platform. The second stage in the algorithm is to construct the beam balance.

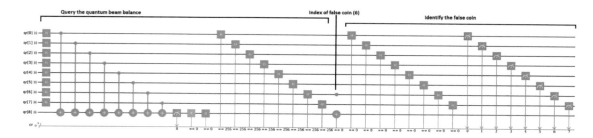

Figure 7-3. *Quantum circuit for the counterfeit coin puzzle with N = 8, k = 1, and fake at index 6 (Note: For full-size viewing, this graph is included in the source code download)*

Step 2: Construct the Quantum Balance

In the previous section, we constructed the superposition of all binary query strings whose Hamming weights are even. In this step, we construct the quantum beam by setting the position of the false coin. Thus given k is the position of the false coin with regard to the binary string |x1, x2, ..., xN>|0>, the quantum beam balance returns

$$|x1, x2, \dots , xN> |0 \oplus xk>$$

This is implemented with a CNOT gate with xk as the control and the second register as the target (see partial Listing 7-2).

Listing 7-2. Construct the Quantum Beam Balance

```
#----- Construct the quantum beam balance
k = indexOfFalseCoin

# Apply the quantum beam balance on the desired superposition state
#(marked by cr equal to zero)
circuit.cx(qr[k], qr[N]).c_if(cr, 0)
```

Step 3: Identify the False Coin

To identify the false coin after querying the balance, apply a Hadamard transform on the binary query string. Assuming that we query the quantum beam balance with binary strings of even Hamming weight, then by performing the measurement in the computational basis after the Hadamard transform, we can identify the false coin because it is the one whose label is different from the majority (see Listing 7-3).

Listing 7-3. Identify the False Coin

```
# --- Identify the false coin
# Apply Hadamard transform on qr[0] ... qr[N-1]
for i in range(N):
  circuit.h(qr[i]).c_if(cr, 0)

# Measure qr[0] ... qr[N-1]
for i in range(N):
  circuit.measure(qr[i], cr[i])

results   = Q_program.execute([circuitName], backend=backend, shots=shots)
answer = results.get_counts(circuitName)

print("Device " + backend + " counts " + str(answer))

# Get most common label
for key in answer.keys():
  normalFlag, _ = Counter(key[1:]).most_common(1)[0]
```

```
for i in range(2,len(key)):
  if key[i] != normalFlag:
    print("False coin index is: ", len(key) - i - 1)
```

When the leftmost bit is 0, the index of the false coin can be determined by finding the one whose values are different from others. For example, for N = 8, false index = 6, then the result should be 010111111 or 001000000. Note that because we use cr[N] to control the operation prior to and after the query to the balance, then

- If the leftmost bit is 0, then we succeed in identifying the false coin.

- If the leftmost bit is 1, we failed to obtain the desired superposition and must repeat the process from the beginning.

Running the program against the remote IBM Q Experience simulator gives the result (under book source Workspace\Ch07\p_counterfeitcoin.py). Note that I am using Windows in this instance:

```
c:\python36-64\python.exe p_counterfeitcoin.py
Device ibmq_qasm_simulator counts {'001000000': 1}
False coin index is:  6
```

If you don't have access to the book source and still want to play with this script, paste the snippets in the previous sections inside the container script in Listing 7-4 (watch out for Python's indentation idiosyncrasies; they will drive you nuts).

Listing 7-4. Counterfeit Coin Puzzle Main Container Script

```
import sys
import matplotlib.pyplot as plt
import numpy as np
from math import pi, cos, acos, sqrt
from collections import Counter
from qiskit import QuantumProgram
sys.path.append('../Config/')
import Qconfig
```

```
# import basic plot tools
from qiskit.tools.visualization import plot_histogram

def main(M = 16, numberOfCoins = 8 , indexOfFalseCoin = 6
  , backend = "local_qasm_simulator" , shots = 1 ):

  if numberOfCoins < 4 or numberOfCoins >= M:
    raise Exception("Please use numberOfCoins between 4 and ", M-1)
  if indexOfFalseCoin < 0 or indexOfFalseCoin >= numberOfCoins:
    raise Exception("indexOfFalseCoin must be between 0 and ",
    numberOfCoins-1)

  // Paste listings 7-1 -> 7-3 here


####################################################
# main
####################################################
if __name__ == '__main__':
  M = 8                          #Maximum qubits available
  numberOfCoins = 4     #Up to M-1, where M is the number of qubits
  available
  indexOfFalseCoin = 2 #This should be 0, 1, ..., numberOfCoins - 1,

  backend   = "ibmq_qasm_simulator"
  #backend = "ibmqx3"
  shots     = 1      # We perform a one-shot experiment

  main(M, numberOfCoins, indexOfFalseCoin, backend, shots)
```

Generalization for Any Number of False Coins

The counterfeit coin puzzle has been generalized by any number of fake coins (k>1) by mathematicians Terhal and Smolin in 1998. Their implementation uses a Balance Oracle model (B-Oracle) such that

- Given an input of N bits $x = x1x2...xn \in \{0, 1\}^N$.

- Construct a query string of N tri-bits such that $q = q1q2... qn \in \{0, 1, -1\}^N$ with the same number of 1s and -1s.

- The answer is 1 bit such that

$$\chi(x;q) = \begin{cases} 0 \text{ if } q1x1 + q2x2 + \dots qnxn = 0 \, (balanced) \\ 1 \, otherwise \qquad\qquad\qquad (tilted) \end{cases}.$$

Tip   An Oracle is the portion of an algorithm regarded as a black box. It is used to simplify circuits and provide complexity comparisons between quantum and classical algorithms. A good Oracle should provide speed, generality, and feasibility.

An example of the B-Oracle in action is shown in Figure 7-4 for two fake coins: k = 2 and N = 6.

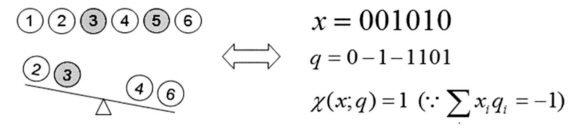

Figure 7-4. *B-Oracle for N = 6 and k = 2)*

All in all, the counterfeit puzzle is the quintessential example of quartic speedup of a quantum algorithm over its classical counterpart. In the next section, we look at another bizarre quasi-magical puzzle called the Mermin-Peres Magic Square.

Mermin-Peres Magic Square

This is another classic puzzle first proposed by physicists David Mermin and A. Peres as an example of quantum pseudo-telepathy or the ability of two players to have some supernatural communication to outside observers. Thanks to the magic of entanglement, this is possible. Let's take a closer look.

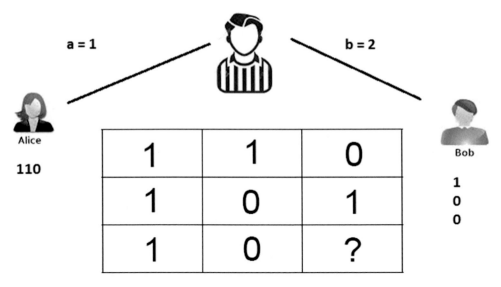

Figure 7-5. *Mermin-Peres Magic Square*

The game starts with two players Alice and Bob against a referee. The magic square is a 3x3 matrix with the following rules (see Figure 7-5):

- All entries are either 0 or 1 such that the sum of entries on each row is even, and the sum of each column is odd. The game is called the magic square because such square is impossible; as shown in Figure 7-5, there is no valid combination where the sum of rows is even and the sum of columns is odd (try it yourself with pen and paper). This is due to the odd number of entries in the matrix.

- The referee sends an integer $a \in \{1,2,3\}$ to Alice and another $b \in \{1,2,3\}$ to Bob. Alice must reply with the a-th row of the square. Bob must reply with the b-th column.

- Alice and Bob win if the sum of Alice's entries is even, the sum of Bob's is odd, and their intersecting answer is the same. Otherwise, the referee wins.

- Prior to the start, Alice and Bob can strategize and share information. For example, they could decide to answer using the matrix in Figure 7-5. However they are not allowed to communicate during the game.

For example, in the preceding matrix, if the referee sends a = 1 to Alice and b = 2 to Bob, Alice would reply with 110 (row 1) and Bob with 100 (column 2). The element in the intersection of the answers (row 1-column 2) is the same (1) so they win the game. It can be shown that in a classical setting, the winning probability for Alice and Bob is at most 8/9. That is, there are eight out of nine permutations in the square for victory. Therefore Alice and Bob's winning probability is at most 88.8%.

Let's put this assertion to the test with a neat exercise to prove that indeed the classical winning probability for the magic square is at most 8/9 (88.88%).

Mermin-Peres Magic Square Exercise

1. Construct a magic square similar to Figure 7-5 using the binary code (1, -1) instead of (1, 0) where the product of the row elements is 1 (even), and the product of the column elements is -1 (odd). Confirm that in fact this is not possible.

2. Create a permutation table for the referee values for a and b using the square in step 1 including

    • A permutation count number.

    • The values for a, b.

    • Alice and Bob's response.

    • The intersection of Alice and Bob's response. Remember that it must be equal for them to win.

    • The result of the game iteration: Win = W, Loose = L.

3. Finally calculate the winning probability and prove that it is at most 8/9. **Note: Answers are at the end of the section.**

Quantum Winning Strategy

Thanks to the power of quantum mechanics and the magic of entanglement, Alice and Bob can do much better. In fact they can win the game 100% of the time, as if they were communicating telepathically, hence the term pseudo-telepathy. A quantum winning strategy was first proposed by Brassard and colleagues[1] and it is divided in three stages:

- *Shared entangled state*: This is the key for Alice and Bob to win 100% of the time.

- *Unitary transformations for Alice and Bob's inputs*: These provide the responses to be sent back to the referee.

- *Measure in the computational basis*: The final stage to construct a final response.

Shared Entangled State

In the quantum winning strategy, Alice and Bob share the entangled state:

$$\Psi = \frac{1}{2}|0011\rangle - \frac{1}{2}|0110\rangle - \frac{1}{2}|1001\rangle + \frac{1}{2}|1100\rangle$$

A circuit implementation requires 2 qubits for Alice and 2 for Bob as shown in Figure 7-6.

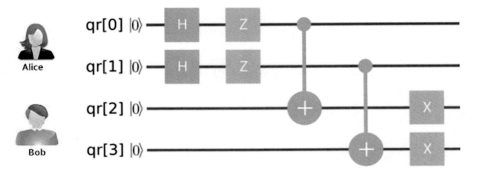

Figure 7-6. *Entangled state for the magic square*

---

[1]Brassard, Broadbent and Tapp. Quantum Pseudo-Telepathy. pp 22, available online at https://arxiv.org/abs/quant-ph/0407221v3.

- We know that the Hadamard maps the basis state

$H|0\rangle \rightarrow \frac{1}{\sqrt{2}}\left(|0\rangle+|1\rangle\right)$. Thus applying for the first 2 qubits yields

$$\Psi = \frac{1}{2}|00\rangle+\frac{1}{2}|01\rangle+\frac{1}{2}|10\rangle+\frac{1}{2}|11\rangle.$$

- Next apply a Z gate to the first 2 qubits. Remember that Z leaves the 0 state unchanged and maps 1 to -1 (flipping the sign of the preceding third term). At this stage the state becomes

$$\Psi = \frac{1}{2}|00\rangle+\frac{1}{2}|01\rangle-\frac{1}{2}|10\rangle+\frac{1}{2}|11\rangle.$$

- Next apply the CNOT gate to entangle qubits 0-2 and 1-3:

$$\Psi = \frac{1}{2}|0000\rangle-\frac{1}{2}\langle0101|-\frac{1}{2}|1010\rangle+\frac{1}{2}|1111\rangle.$$

- Finally flip the last 2 qubits with the X gate for

$$\Psi = \frac{1}{2}|0011\rangle-\frac{1}{2}|0110\rangle-\frac{1}{2}|1001\rangle+\frac{1}{2}|1100\rangle.$$

The Python script to construct the entangled state is given in Listing 7-5.

Listing 7-5. Quantum Winning Strategy Entangled State

```
# Create the entangle state
Q_program = QuantumProgram()
Q_program.set_api(Qconfig.APItoken, Qconfig.config["url"])

# 4 qubits (Alice = 2, Bob = 2)
N = 4

# Creating registers
qr = Q_program.create_quantum_register("qr", N)

# for recording the measurement on qr
cr = Q_program.create_classical_register("cr", N)

circuitName = "sharedEntangled"
sharedEntangled = Q_program.create_circuit(circuitName, [qr], [cr])

#Create uniform superposition of all strings of length 2
for i in range(2):
    sharedEntangled.h(qr[i])
```

```
#The amplitude is minus if there are odd number of 1s
for i in range(2):
      sharedEntangled.z(qr[i])
```

```
#Copy the content of the first two qubits to the last two qubits
for i in range(2):
      sharedEntangled.cx(qr[i], qr[i+2])
```

```
#Flip the last two qubits
for i in range(2,4):
      sharedEntangled.x(qr[i])
```

   With the shared entangled state, Alice and Bob can now start the game and receive their inputs from the referee.

Unitary Transformations

Upon receiving their inputs $a \in \{1,2,3\}$ and $b \in \{1,2,3\}$, Alice and Bob apply the following unitary transformations: A1, A2, A3 for Alice and B1, B2, B3 for Bob to the shared entangled states:

$$A1 = \frac{1}{\sqrt{2}}\begin{bmatrix} i & 0 & 0 & 1 \\ 0 & -i & 1 & 0 \\ 0 & i & 1 & 0 \\ 1 & 0 & 0 & i \end{bmatrix}, A2 = \frac{1}{2}\begin{bmatrix} i & 1 & 1 & i \\ -i & 1 & -1 & i \\ i & 1 & 1 & -i \\ -i & 1 & -1 & -i \end{bmatrix}, A3 = \frac{1}{2}\begin{bmatrix} -1 & -1 & -1 & 1 \\ 1 & 1 & -1 & 1 \\ 1 & -1 & 1 & 1 \\ 1 & -1 & -1 & -1 \end{bmatrix}$$

$$B1 = \frac{1}{2}\begin{bmatrix} i & -i & 1 & 1 \\ -i & -i & 1 & -1 \\ 1 & 1 & -i & i \\ -i & i & 1 & 1 \end{bmatrix}, B2 = \frac{1}{2}\begin{bmatrix} -1 & i & 1 & i \\ 1 & i & 1 & -i \\ 1 & -i & 1 & i \\ -1 & -i & 1 & -i \end{bmatrix}, B3 = \frac{1}{\sqrt{2}}\begin{bmatrix} 1 & 0 & 0 & 1 \\ -1 & 0 & 0 & 1 \\ 0 & 1 & 1 & 0 \\ 0 & 1 & -1 & 0 \end{bmatrix}$$

Note   Remember that by applying the preceding transformations to their entangled states, Alice and Bob are able to construct the first 2 bits of their respective responses to the referee.

Listing 7-6 shows the unitary transformations for Alice and Bob with equivalent graphical circuits in Table 7-1.

Listing 7-6. Unitary Transformations for Alice and Bob

```python
#------  circuits of Alice's and Bob's operations.
#we first define controlled-u gates required to assign phases
from math import pi

def ch(qProg, a, b):
  """ Controlled-Hadamard gate """
  qProg.h(b)
  qProg.sdg(b)
  qProg.cx(a, b)
  qProg.h(b)
  qProg.t(b)
  qProg.cx(a, b)
  qProg.t(b)
  qProg.h(b)
  qProg.s(b)
  qProg.x(b)
  qProg.s(a)
  return qProg

def cu1pi2(qProg, c, t):
  """ Controlled-u1(phi/2) gate """
  qProg.u1(pi/4.0, c)
  qProg.cx(c, t)
  qProg.u1(-pi/4.0, t)
  qProg.cx(c, t)
  qProg.u1(pi/4.0, t)
  return qProg

def cu3pi2(qProg, c, t):
  """ Controlled-u3(pi/2, -pi/2, pi/2) gate """
  qProg.u1(pi/2.0, t)
  qProg.cx(c, t)
```

```
  qProg.u3(-pi/4.0, 0, 0, t)
  qProg.cx(c, t)
  qProg.u3(pi/4.0, -pi/2.0, 0, t)
  return qProg

#---------------------------------------------------------------------
# Define circuits used by Alice and Bob for each of their inputs: 1,2,3
# dictionary for Alice's operations/circuits
aliceCircuits = {}

# Quantum circuits for Alice 1, 2, 3
for idx in range(1, 4):
  circuitName = "Alice"+str(idx)
  aliceCircuits[circuitName]
    = Q_program.create_circuit(circuitName, [qr], [cr])
  theCircuit = aliceCircuits[circuitName]

  if idx == 1:
    #the circuit of A_1
    theCircuit.x(qr[1])
    theCircuit.cx(qr[1], qr[0])
    theCircuit = cu1pi2(theCircuit, qr[1], qr[0])
    theCircuit.x(qr[0])
    theCircuit.x(qr[1])
    theCircuit = cu1pi2(theCircuit, qr[0], qr[1])
    theCircuit.x(qr[0])
    theCircuit = cu1pi2(theCircuit, qr[0], qr[1])
    theCircuit = cu3pi2(theCircuit, qr[0], qr[1])
    theCircuit.x(qr[0])
    theCircuit = ch(theCircuit, qr[0], qr[1])
    theCircuit.x(qr[0])
    theCircuit.x(qr[1])
    theCircuit.cx(qr[1], qr[0])
    theCircuit.x(qr[1])
```

```
  elif idx == 2:
    theCircuit.x(qr[0])
    theCircuit.x(qr[1])
    theCircuit = cu1pi2(theCircuit, qr[0], qr[1])
    theCircuit.x(qr[0])
    theCircuit.x(qr[1])
    theCircuit = cu1pi2(theCircuit, qr[0], qr[1])
    theCircuit.x(qr[0])
    theCircuit.h(qr[0])
    theCircuit.h(qr[1])

  elif idx == 3:
    theCircuit.cz(qr[0], qr[1])
    theCircuit.swap(qr[0], qr[1]) # not supported in composer
    theCircuit.h(qr[0])
    theCircuit.h(qr[1])
    theCircuit.x(qr[0])
    theCircuit.x(qr[1])
    theCircuit.cz(qr[0], qr[1])
    theCircuit.x(qr[0])
    theCircuit.x(qr[1])

  #measure the first two qubits in the computational basis
  theCircuit.measure(qr[0], cr[0])
  theCircuit.measure(qr[1], cr[1])

# dictionary for Bob's operations/circuits
bobCircuits = {}

# Quantum circuits for Bob when receiving 1, 2, 3
for idx in range(1,4):
  circuitName = "Bob"+str(idx)
  bobCircuits[circuitName]
    = Q_program.create_circuit(circuitName, [qr], [cr])
  theCircuit = bobCircuits[circuitName]
  if idx == 1:
    theCircuit.x(qr[2])
    theCircuit.x(qr[3])
```

```
            theCircuit.cz(qr[2], qr[3])
            theCircuit.x(qr[3])
            theCircuit.u1(pi/2.0, qr[2])
            theCircuit.x(qr[2])
            theCircuit.z(qr[2])
            theCircuit.cx(qr[2], qr[3])
            theCircuit.cx(qr[3], qr[2])
            theCircuit.h(qr[2])
            theCircuit.h(qr[3])
            theCircuit.x(qr[3])
            theCircuit = cu1pi2(theCircuit, qr[2], qr[3])
            theCircuit.x(qr[2])
            theCircuit.cz(qr[2], qr[3])
            theCircuit.x(qr[2])
            theCircuit.x(qr[3])

        elif idx == 2:
            theCircuit.x(qr[2])
            theCircuit.x(qr[3])
            theCircuit.cz(qr[2], qr[3])
            theCircuit.x(qr[3])
            theCircuit.u1(pi/2.0, qr[3])
            theCircuit.cx(qr[2], qr[3])
            theCircuit.h(qr[2])
            theCircuit.h(qr[3])

        elif idx == 3:
            theCircuit.cx(qr[3], qr[2])
            theCircuit.x(qr[3])
            theCircuit.h(qr[3])

    #measure the third and fourth qubits in the computational basis
    theCircuit.measure(qr[2], cr[2])
    theCircuit.measure(qr[3], cr[3])
```

Table 7-1 shows quantum circuits for the unitary transformations A1-3, B1-3 for IBM Q Experience Composer.

Table 7-1. Quantum Circuits for the Unitary Transformations in Listing 7-6

Transformation	Circuit
$A1 = \dfrac{1}{\sqrt{2}}\begin{bmatrix} i & 0 & 0 & 1 \\ 0 & -i & 1 & 0 \\ 0 & i & 1 & 0 \\ 1 & 0 & 0 & i \end{bmatrix}$	
$A2 = \dfrac{1}{2}\begin{bmatrix} i & 1 & 1 & i \\ -i & 1 & -1 & i \\ i & 1 & 1 & -i \\ -i & 1 & -1 & -i \end{bmatrix}$	
$B1 = \dfrac{1}{2}\begin{bmatrix} i & -i & 1 & 1 \\ -i & -i & 1 & -1 \\ 1 & 1 & -i & i \\ -i & i & 1 & 1 \end{bmatrix}$	

(continued)

Table 7-1. (continued)

Transformation	Circuit
$B2 = \dfrac{1}{2}\begin{bmatrix} -1 & i & 1 & i \\ 1 & i & 1 & -i \\ 1 & -i & 1 & i \\ -1 & -i & 1 & -i \end{bmatrix}$	
$B3 = \dfrac{1}{\sqrt{2}}\begin{bmatrix} 1 & 0 & 0 & 1 \\ -1 & 0 & 0 & 1 \\ 0 & 1 & 1 & 0 \\ 0 & 1 & -1 & 0 \end{bmatrix}$	

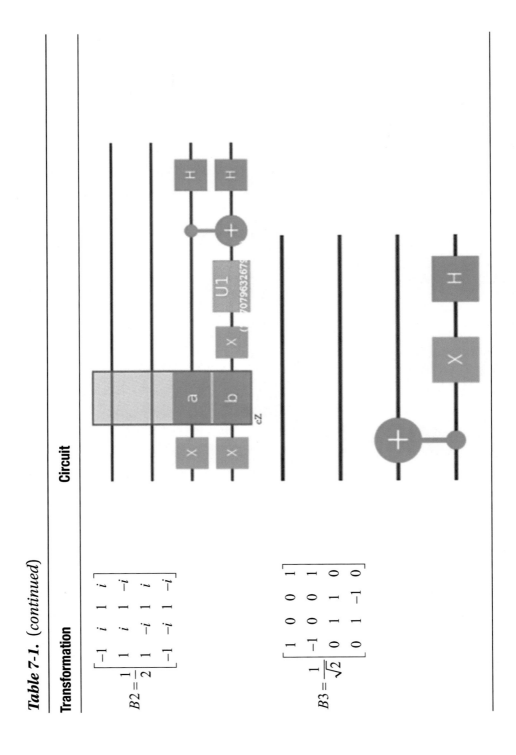

In Table 7-1 note that A3 is not included due to the fact that the Composer does not support the swap gate required by Listing 7-6. This does not mean the quantum program can't be run in the simulator or real device however. It simply means the circuit cannot be created in the Composer. Thus for the final step, Alice and Bob measure their qubits in the computational basis.

Measure in the Computational Basis

After measurement, Alice and Bob end up with 2 bits each which represent their respective outputs. To obtain the third bit, and thus a final answer, they apply their parity rules. That is, Alice's sum must be even, and Bob's must be odd. For example, for $a = 2$, $b = 3$ (see Table 7-2)

$$(A2 \otimes B3)|\psi\rangle = \frac{1}{2\sqrt{2}}[|0000\rangle - |0010\rangle - |0101\rangle + |0111\rangle + |1001\rangle$$
$$- |1011\rangle - |1100\rangle - |1110\rangle$$

Table 7-2. *Answer Permutations for a = 2, b =3 of the Magic Square*

ψ	Alice's answer	Bob's answer	Square
|0000>	000	001	$\begin{bmatrix} & 0 & \\ 0 & 0 & 0 \\ & 1 & \end{bmatrix}$
|0010>	000	100	$\begin{bmatrix} & & 1 \\ 0 & 0 & 0 \\ & & 0 \end{bmatrix}$
|0101>	011	010	$\begin{bmatrix} & & 1 \\ 0 & 1 & 1 \\ & & 0 \end{bmatrix}$

(continued)

Table 7-2. (*continued*)

ψ	Alice's answer	Bob's answer	Square
|0111>	011	111	$\begin{bmatrix} & & 1 \\ 0 & 1 & 1 \\ & & 1 \end{bmatrix}$
|1001>	101	010	$\begin{bmatrix} & & 0 \\ 1 & 0 & 1 \\ & & 0 \end{bmatrix}$
|1011>	101	111	$\begin{bmatrix} & & 1 \\ 1 & 0 & 1 \\ & & 1 \end{bmatrix}$
|1100>	110	001	$\begin{bmatrix} & & 0 \\ 1 & 1 & 0 \\ & & 1 \end{bmatrix}$
|1110>	110	101	$\begin{bmatrix} & & 1 \\ 1 & 1 & 0 \\ & & 1 \end{bmatrix}$

Listing 7-7 shows a section of the script to loop through all the rounds of the magic square:

- It loops through *a[1,3]* and *b[1,3]* inclusive.

- For each (a, b), a circuit for Alice (Alice-a) and a circuit for Bob (Bob-b) are retrieved from Listing 7-6.

- The shared entangled state ψ and Alice-a and Bob-b circuits are submitted for execution to the simulator or real quantum device.

- Two bits are extracted for Alice and two for Bob from the answer such as {'0011': 1}.

- The parity rules are applied: Alice's sum must be even, and Bob's sum must be odd.

- Finally the answer is verified, and the winning probability is displayed.

Listing 7-7. Script for All Rounds of the Magic Square

```
def all_rounds(backend, real_dev, shots=10):
  nWins = 0
  nLost = 0
  for a in range(1,4):
    for b in range(1,4):
      print("Asking Alice and Bob with a and b are: ", a,b)
      rWins = 0
      rLost = 0

      aliceCircuit   = aliceCircuits["Alice" + str(a)]
      bobCircuit     = bobCircuits["Bob" + str(b)]
      circuitName    = "Alice" + str(a) + "Bob"+str(b)
      Q_program.add_circuit(circuitName, sharedEntangled+aliceCircuit+bobCi
      rcuit)

      if real_dev:
        ibmqx2_backend = Q_program.get_backend_configuration(backend)
        ibmqx2_coupling = ibmqx2_backend['coupling_map']
        results = Q_program.execute([circuitName], backend=backend,
        shots=shots
              , coupling_map=ibmqx2_coupling, max_credits=3, wait=10,
              timeout=240)
      else:
        results = Q_program.execute([circuitName], backend=backend,
        shots=shots)

      answer = results.get_counts(circuitName)

      for key in answer.keys():
        kfreq = answer[key] #frequencies of keys obtained from measurements
        aliceAnswer = [int(key[-1]), int(key[-2])]
```

```
    bobAnswer    = [int(key[-3]), int(key[-4])]
    if sum(aliceAnswer) % 2 == 0:
      aliceAnswer.append(0)
    else:
      aliceAnswer.append(1)
    if sum(bobAnswer) % 2 == 1:
      bobAnswer.append(0)
    else:
      bobAnswer.append(1)

    if(aliceAnswer[b-1] != bobAnswer[a-1]):
      #print(a, b, "Alice and Bob lost")
      nLost += kfreq
      rLost += kfreq
    else:
      #print(a, b, "Alice and Bob won")
      nWins += kfreq
      rWins += kfreq
   print("\t#wins = ", rWins, "out of ", shots, "shots")

  print("Number of Games = ", nWins+nLost)
  print("Number of Wins = ", nWins)
  print("Winning probabilities = ", (nWins*100.0)/(nWins+nLost))

################################################
# main
################################################
if __name__ == '__main__':
  backend = "ibmq_qasm_simulator"
  #backend = "ibmqx2"
  real_dev = False

  all_rounds(backend, real_dev)
```

A run of Listing 7-7 against the IBM Q Experience remote simulator is shown in Listing 7-8.

Listing 7-8. Simplified Standard Output from a Run of All Rounds of the Magic Square

```
c:\python36-64\python.exe p_magicsq.py
For a = 1 , b = 1
ibmq_qasm_simulator answer: 1000 Alice: [0, 0, 0] Bob:[0, 1, 0]
ibmq_qasm_simulator answer: 1010 Alice: [0, 1, 1] Bob:[0, 1, 0]
ibmq_qasm_simulator answer: 1111 Alice: [1, 1, 0] Bob:[1, 1, 1]
ibmq_qasm_simulator answer: 0111 Alice: [1, 1, 0] Bob:[1, 0, 0]
ibmq_qasm_simulator answer: 0000 Alice: [0, 0, 0] Bob:[0, 0, 1]
ibmq_qasm_simulator answer: 0101 Alice: [1, 0, 1] Bob:[1, 0, 0]
        #wins =  10 out of  10 shots
For a = 1 , b = 2
ibmq_qasm_simulator answer: 1000 Alice: [0, 0, 0] Bob:[0, 1, 0]
ibmq_qasm_simulator answer: 1001 Alice: [1, 0, 1] Bob:[0, 1, 0]
ibmq_qasm_simulator answer: 1111 Alice: [1, 1, 0] Bob:[1, 1, 1]
ibmq_qasm_simulator answer: 0110 Alice: [0, 1, 1] Bob:[1, 0, 0]
ibmq_qasm_simulator answer: 0000 Alice: [0, 0, 0] Bob:[0, 0, 1]
ibmq_qasm_simulator answer: 0001 Alice: [1, 0, 1] Bob:[0, 0, 1]
        #wins =  10 out of  10 shots
...
For a = 3 , b = 3
ibmq_qasm_simulator answer: 1000 Alice: [0, 0, 0] Bob:[0, 1, 0]
ibmq_qasm_simulator answer: 1011 Alice: [1, 1, 0] Bob:[0, 1, 0]
ibmq_qasm_simulator answer: 1101 Alice: [1, 0, 1] Bob:[1, 1, 1]
ibmq_qasm_simulator answer: 1110 Alice: [0, 1, 1] Bob:[1, 1, 1]
ibmq_qasm_simulator answer: 0111 Alice: [1, 1, 0] Bob:[1, 0, 0]
ibmq_qasm_simulator answer: 0010 Alice: [0, 1, 1] Bob:[0, 0, 1]
        #wins =  10 out of  10 shots
Number of Games =  90
Number of Wins =  90
Winning probability =  100.0
```

Note   If running in a real device, the winning probability will not be 100% due to environmental noise and gate error.

Answers for the Mermin-Peres Magic Square Exercise

1. A magic square whose row product is even and whose column product is odd is given in the following presentation. Note that such square is not possible due to the odd number of cells.

-1	-1	1
-1	1	-1
-1	1	?

2. The permutation table for the square in answer 1 is

N	a	b	Alice	Bob	Intersection	Win/Loose
1	1	1	-1,-1,1	-1,-1,-1	-1/-1	W
2	1	2	-1,-1,1	-1,1,1	-1/-1	W
3	1	3	-1,-1,1	1,-1,? (1)	1/1	W
4	2	1	-1,1,-1	-1,-1,-1	-1/-1	W
5	2	2	-1,1,-1	-1,1,1	1/1	W
6	2	3	-1,1,-1	1,-1,? (1)	-1/-1	W
7	3	1	-1,1,? (-1)	-1,-1,-1	-1/-1	W
8	3	2	-1,1,? (-1)	-1,1,1	1/1	W
9	3	3	-1,1,? (-1)	1,-1,? (1)	-1/1	L

3. Note that, in the previous step rows 7-9, Alice's answer must be
-1 so the product can be even (1). Plus, in columns 3, 6, and 9,
Bob's answer must be 1 so his product can be odd (−1). Finally, the
probability is calculated by dividing the total number of wins by
the total number of permutations. Thus

$$P = \frac{\sum W}{N} = \frac{8}{9} = 88.88\%$$

In this chapter you have learned how the power of quantum entanglement can
provide significant speedups over classical computation. With a quantum beam
balance, it is possible to achieve quartic speedups for classical puzzles like the
counterfeit coin problem. For others, such as the magic square, entanglement gives
a quasi-magical telepathy among players. Now, if only Brassard and colleagues
could come up with a quantum winning strategy for Black Jack or Poker, we will
all be making a killing in Vegas right now. All in all, this chapter has shown how
quantum mechanics is as confusing, bizarre, and fascinating as always. It never
fails to deliver.

In the next and final chapter, you will learn about arguably the most famous
quantum algorithm of them all: the notorious Shor's integer factorization. An algorithm
that may crumble asymmetric cryptography!

CHAPTER 8

Faster Search plus Threatening the Foundation of Asymmetric Cryptography with Grover and Shor

This chapter brings proceedings to a close with two algorithms that have generated excitement about the possibilities of practical quantum computation:

- *Grover's search*: This is an unstructured quantum search algorithm created by Lov Grover which is capable of finding an input with high probability using a black box function or Oracle. It can find an item in $O(\sqrt{N})$ steps as opposed to a classical average of N/2 steps.

- *Shor's integer factorization*: The notorious quantum factorization that experts say could bring current asymmetric cryptography to its knees. Shor can factorize integers in approximately $\log(n^3)$ steps as opposed to the fastest classical algorithm, the Number Field Sieve at $\exp\left(k * \log\left(n^{\frac{1}{3}}\right)(\log\log n)^{\frac{2}{3}}\right)$.

Let's get started.

© Vladimir Silva 2018
V. Silva, *Practical Quantum Computing for Developers*, https://doi.org/10.1007/978-1-4842-4218-6_8

Quantum Unstructured Search

Grover's algorithm is an unstructured search quantum procedure to find an entry of
n bits on a digital haystack of N elements. As shown in Figure 8-1, Grover's quantum
algorithm provides significant speedup at $O\left(\sqrt{N}\right)$ steps. It may not seem much
compared to the classical solution, but when we are talking about millions of entries,
then the square root of 10^6 is much faster than 10^6.

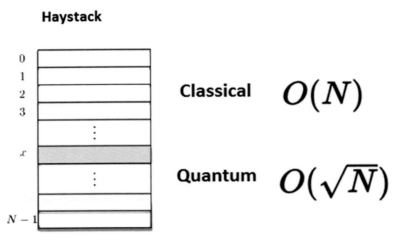

Figure 8-1. *Unstructured search time complexities*

If x is the element we are looking for, then Grover's algorithm can be described by the
following pseudocode:

1. Prepare the input given f: {0, 1, ... , N-1} → {0,1}. Note that the
    size of the input is 2^n where n is the number of bits and N is the
    number of steps or size of the haystack. The ultimate goal is to find
    x such that f(x) = 1.

2. Apply a basis superposition to all qubits in the input.

3. Perform a phase inversion on the input qubits.

4. Perform an inversion about the mean on the input.

5. Repeat steps 3 and 4 at least $\sqrt{N}$ times. There is a high probability
    that x will be found at this point.

Let's take a closer look at the critical phase inversion and inversion about the mean
steps.

314

Phase Inversion

This is the first step in the algorithm and must be performed in a superposition of all states in the haystack. If the element we are looking for is x' where $f(x') = 1$, then the superposition can be expressed as $\sum \alpha \,|\, x\rangle$. Ultimately, what phase inversion does is

$$\sum \alpha \,|x\rangle \xrightarrow{\text{\textit{Phase Inversion}}} \begin{cases} \sum \alpha \,|x\rangle \; \textit{if } x \neq x' \\ -\alpha \,|x'\rangle \, \textit{Otherwise} \end{cases}$$

That is, if a given x is not the element we are looking for ($x \neq x'$), then it leaves the superposition intact; otherwise it inverts the phase (the sign of the complex coefficient α of the qubit – see Figure 8-2 for a pictorial representation).

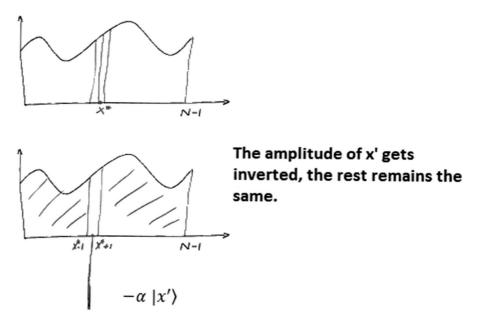

The amplitude of x' gets inverted, the rest remains the same.

Figure 8-2. *Pictorial representation of phase inversion*

This is the first step in Grover's algorithm; we'll see how phase inversion helps on finding the element we are looking for, but for now let's look at the second step: inversion about the mean.

Inversion About the Mean

Given the previous superposition $\sum \alpha \mid x\rangle$, we define the mean μ, as the average value of the amplitudes:

$$\mu = \frac{\sum_{x=0}^{N-1} \alpha_x}{N}$$

Now we must flip the amplitudes about this mean. That is,

$$\alpha_x \rightarrow \left(2\mu - \alpha_x \right)$$

$$\sum \alpha_x \mid x\rangle \rightarrow \sum \left(2\mu - \alpha_x \right) \mid x\rangle$$

To better understand this, Figure 8-3 shows a pictorial representation of inversion about the mean.

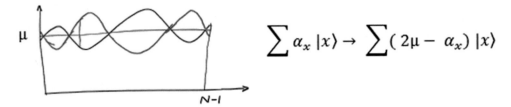

$$\mu \qquad\qquad\qquad\qquad\qquad\qquad \sum \alpha_x \mid x\rangle \rightarrow \sum \left(2\mu - \alpha_x \right) \mid x\rangle$$

$$N-1$$

Figure 8-3. *Graphical representation of inversion about the mean*

Figure 8-3 shows the superimposed state of the qubits as defined by the wave function ψ. The mean or μ of this function is shown as the horizontal line in the chart. What inversion about the mean does is it mirrors the wave function ψ over the mean μ resulting in a mirror wave (shown with a dotted line). This is equivalent to rotating the waves over the axis μ. Let's make sense of all this by putting all steps together to see them in action:

1. Superposition

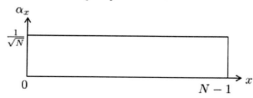

2. Phase Inversion

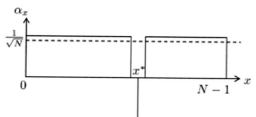

3. Inversion about the mean

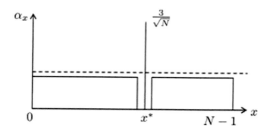

Amplitude increase after $\sqrt{N}$ steps

Figure 8-4. *Single Grover's iteration*

In Figure 8-4:

- The superposition of all qubits puts all amplitudes at $\dfrac{1}{\sqrt{N}}$.

- Next, a phase inversion puts the amplitude for x′ at $-\dfrac{1}{\sqrt{N}}$. Note that this has the effect of lowering slightly the value of the mean μ, as shown by the dotted line in Figure 8-4 step 2.

- After the inversion about the mean, the mean amplitude drops a little bit, but x′ goes way high, as much as $\dfrac{2}{\sqrt{N}}$ above the mean μ.

- If we repeat this sequence, the amplitude of x′ increases by about $\dfrac{2}{\sqrt{N}}$ until that, in about $\sqrt{N}$ steps, the amplitude becomes $\dfrac{1}{\sqrt{2}}$.

- At this point, if we measure our qubits, the probability of finding x′ (the element we are looking for), as defined by quantum mechanics, is the square of the amplitude. That is, ½.

- Thus we are done. In roughly $\sqrt{N}$ steps, we have found the marked element x′.

Now, let's put all this together in a quantum circuit and corresponding code implementation.

Practical Implementation

We'll take a look at a circuit for Grover's algorithm in IBM Q Experience. The circuit demonstrates a single iteration of the algorithm using 2 qubits as shown in Figure 8-5.

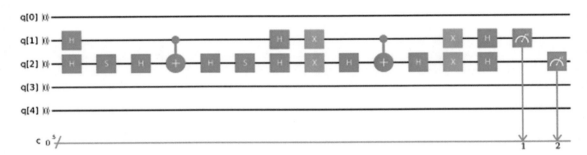

Figure 8-5. *Quantum circuit for Grover's algorithm with 2 qubits and A = 01*

A Python script that creates the circuit in Figure 8-5 is shown in Listing 8-1.

Listing 8-1. Python Script for Circuit in Figure 8-5

```
import sys,time,math

# Importing QISKit
from qiskit import QuantumCircuit, QuantumProgram

# Q Experience config
sys.path.append('../Config/')
import Qconfig

# Import basic plotting tools
from qiskit.tools.visualization import plot_histogram

# Set the input bits to search for
def input_phase (circuit, qubits):
    # Uncomment for A = 00
    # Comment for A = 11
```

```
  circuit.s(qubits[0])
  #circuit.s(qubits[1])
  return

# circuit: Grover 2-qubit circuit
# qubits: Array of qubits (size 2)
def invert_over_the_mean (circuit, qubits):
  for i in range (2):
    circuit.h(qubits[i])
    circuit.x(qubits[i])

  circuit.h(qubits[1])
  circuit.cx(qubits[0], qubits[1])
  circuit.h(qubits[1])

  for i in range (2):
    circuit.x(qubits[i])
    circuit.h(qubits[i])

def invert_phase (circuit, qubits):
  # Oracle
  circuit.h(qubits[1])
  circuit.cx(qubits[0], qubits[1])
  circuit.h(qubits[1])

def main():
  # Quantum program setup
  qp = QuantumProgram()

  qp.set_api(Qconfig.APItoken, Qconfig.config["url"])

  # Create qubits/registers
  size = 2
  q = qp.create_quantum_register('q', size)
  c = qp.create_classical_register('c', size)

  # Quantum circuit
  grover = qp.create_circuit('grover', [q], [c])
```

```python
  # 1. put all qubits in superposition
  for i in range (size):
    grover.h(q[i])

  # Set the input
  input_phase(grover, q)

  # 2. Phase inversion
  invert_phase(grover, q)

  input_phase(grover, q)

  # 3. Invert over the mean
  invert_over_the_mean (grover, q)

  # measure
  for i in range (size):
    grover.measure(q[i], c[i])

  circuits = ['grover']

  # Execute the quantum circuits on the simulator
  backend = "local_qasm_simulator"
  # the number of shots in the experiment
  shots = 1024

  result = qp.execute(circuits, backend=backend, shots=shots
          , max_credits=3, timeout=240)
  counts = result.get_counts("grover")
  print("Counts:" + str(counts))

  # Optional
  #plot_histogram(counts)

#############################################
# main
if __name__ == '__main__':
  start_time = time.time()
  main()
  print("--- %s seconds ---" % (time.time() - start_time))
```

- Listing 8-1 performs a single interaction of Grover's algorithm for a 2-bit input using 2 qubits. Even though the pseudocode in the previous section states that the total number of iterations is given by roughly $\sqrt{N}$ steps, the inversion about the mean requires this value to be multiplied by $\pi/4$ and its *floor* extracted (see the proof next to Figure 8-8). Therefore, we end up with $IT = floor\left(\sqrt{N} * \dfrac{\pi}{4}\right)$ where

  $N = 2^{bits}$. Thus, for 2 bits we get $IT = floor\left(\sqrt{4} * \dfrac{\pi}{4}\right) = floor(1.57) = 1$.

- The script begins by creating a quantum circuit with 2 qubits and two classical registers to store their measurements.

- Next, all qubits are put in superposition using the Hadamard gate.

- Before the iteration, the input is prepared using the phase gate (S) and the rules in Table 8-1.

Table 8-1. *Input Preparation Rules for Listing 8-1*

Input (A)	Gates/qubits
00	S(01)
10	S(0)
01	S(1)
11	None

- Next, perform a phase inversion followed by an inversion about the mean on the input qubits corresponding to a single iteration of the algorithm.

- Finally, measure the results and execute the circuit in the local or remote simulator. Print the result counts.

Generalized Circuit

In broad terms, the circuit in Figure 8-5 can be generalized to any number of input qubits as shown in Figure 8-6.

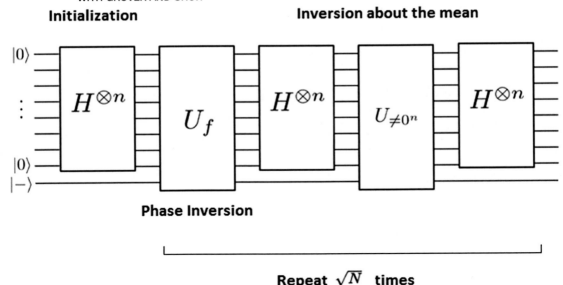

Figure 8-6. *Generalization of Grover's algorithm for an arbitrary number of qubits*

- The first box in Figure 8-6 puts all qubits in superposition by applying the Hadamard gate to the input of size n. This is the initialization step.

- Next, the phase inversion circuit (U_f) receives the superimposed input $\psi = \sum \alpha \,|\, x \rangle$ and a phase input (minus state). This has the desired effect of putting the phase exactly where we want it. Thus the output becomes $\sum \alpha \,(-1)^{f(x)} \,|\, x \rangle$. But how can this be achieved? The answer is that, by applying an exclusive OR on the minus state input, we obtain the desired effect $|b\rangle \rightarrow |\,f(x) \oplus b \rangle$ as shown in Figure 8-7. The third row of the XOR truth table between f(x) and b (the right side of Figure 8-7) shows the phase inversion effect.

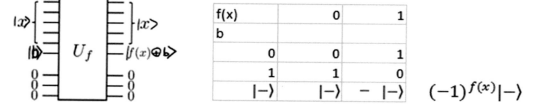

f(x)	0	1
b		
0	0	1
1	1	0
$\lvert-\rangle$	$\lvert-\rangle$	$-\lvert-\rangle$

$(-1)^{f(x)}\lvert-\rangle$

Figure 8-7. *Phase inversion circuit*

- Finally, as shown in Figure 8-3, inversion about the mean is the same as doing the reflection about $\lvert\mu\rangle = 1/\sqrt{N}\sum_x \lvert x\rangle$. To better visualize this, the superimposed state ψ and the mean μ can be represented as vectors over a 2D space as shown in Figure 8-8. To reflect ψ, create an orthogonal vector to μ, then project ψ over the new quadrant at the same angle θ.

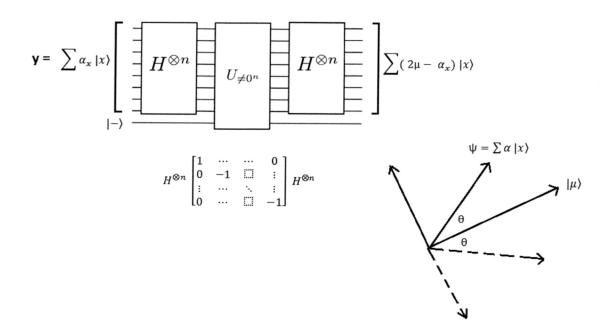

Figure 8-8. *Inversion over the mean circuit*

The proof that inversion over the mean transforms $\sum \alpha_x \mid x\rangle \rightarrow \sum (2\mu - \alpha_x) \mid x\rangle$ involves three steps, as shown by the circuit in Figure 8-8.

1. Transform $|\mu\rangle$ to the all zeros vector $|0, ..., 0\rangle$. This is achieved by applying the Hadamard gate to the input.

2. Reflect about the all zeros vector $|0, ..., 0\rangle$. This can be done by

$$\begin{bmatrix} 1 & \cdots & \cdots & 0 \\ 0 & -1 & & \vdots \\ \vdots & \cdots & \ddots & \vdots \\ 0 & \cdots & & -1 \end{bmatrix}$$

multiplying it by the sparse matrix

3. Transform $|0, ..., 0\rangle$ back to $|\mu\rangle$ by applying the Hadamard again.

Thus

$$H^{\otimes n} \begin{bmatrix} 1 & \cdots & \cdots & 0 \\ 0 & -1 & & \vdots \\ \vdots & \cdots & \ddots & \vdots \\ 0 & \cdots & & -1 \end{bmatrix} H^{\otimes n} = H^{\otimes n} \left(\begin{bmatrix} 2 & \cdots & 0 \\ \vdots & \ddots & \vdots \\ 0 & \cdots & 0 \end{bmatrix} - I \right) H^{\otimes n} = H^{\otimes n} \begin{bmatrix} 2 & \cdots & 0 \\ \vdots & \ddots & \vdots \\ 0 & \cdots & 0 \end{bmatrix} H^{\otimes n} - H^{\otimes n} I H^{\otimes n}$$

$$= \begin{bmatrix} 2/N & \cdots & 2/N \\ \vdots & \ddots & \vdots \\ 2/N & \cdots & 2/N \end{bmatrix} - I = \begin{bmatrix} \frac{2}{N}-1 & \cdots & 2/N \\ \vdots & \ddots & \vdots \\ 2/N & \cdots & \frac{2}{N}-1 \end{bmatrix} \qquad (1)$$

Note that $H^{\otimes n} I H^{\otimes n} = I$ and $H = \frac{2}{\sqrt{N}} \mid x\rangle$. Finally, applying matrix (1) to the state $\psi = \alpha_x \mid x\rangle$ yields

$$\begin{bmatrix} \frac{2}{N}-1 & \cdots & 2/N \\ \vdots & \ddots & \vdots \\ 2/N & \cdots & \frac{2}{N}-1 \end{bmatrix} \begin{bmatrix} \alpha_0 \\ \vdots \\ \alpha_x \\ \vdots \\ \alpha_{N-1} \end{bmatrix} \rightarrow \begin{bmatrix} \vdots \\ 2/N\sum\alpha_y - \alpha_x \\ \vdots \end{bmatrix} = 2\mu - \alpha_x \ where \ 2/N\sum\alpha_y = 2\mu$$

So this is Grover's algorithm for unstructured search. It is fast, powerful, and soon to be hard at work on the data center cranking up all kinds of database searches. Given its significant performance boost over its classical cousin, chances are that in a few years, when quantum computers become more business friendly, most web searches will be performed by this quantum powerhouse. Before we finish, it is worth noting that, by the time of this writing, a useful implementation or experiment (one that can find a real thing) does not exist for IBM Q Experience. Hopefully this will change in the future, but for now Grover's algorithm lives in the theoretical side of things. In the next section, we close strong by looking at the famous Shor's algorithm for integer factorization.

Integer Factorization with Shor's Algorithm

The game of cat and mouse between cryptography and crypto analysis rages on: the first, devising new ways to encrypt our everyday data, and the latter probing for weaknesses, always looking for a crack to bring it down. Current asymmetric cryptography relies on the well-known difficulty of factoring very large primes (in the hundreds of digits range). This section looks at the inner workings of Shor's algorithm, a method that gives exponential speedup for integer factorization using a quantum computer. This is followed by an implementation using a library called ProjectQ. Next, we simulate for sample integers and evaluate the results. Finally we look at current and future directions of integer factorization in quantum systems. Let's get started.

Challenging Asymmetric Cryptography with Quantum Factorization

In the pivotal paper "Polynomial-Time Algorithms for Prime Factorization and Discrete Logarithms on a Quantum Computer,"[1] Peter Shor proposed a quantum factorization method using a principle known to mathematicians for a long time: find the period (also known as order) of an element a in the multiplicative group modulo N, that is, the least positive integer such that

$$x^r \equiv 1 \pmod{N}$$

where N is the number to factor and r is the period of x modulo N.

---

[1]Peter Shor. Polynomial-Time Algorithms for Prime Factorization and Discrete Logarithms on a
Quantum Computer.

Tip   Large integer factorization is a problem that has puzzled mathematicians for millennia. In 1976, G. L. Miller postulated that using randomization, factorization can be reduced to finding the period of an element **a modulo N**, thus greatly simplifying this puzzle. This is the basic idea behind Shor's algorithm.

Shor divided his algorithm in three stages, two of which are performed by a classical computer in polynomial time:

1. *Input preparation*: Done in a classical computer in polynomial time `log (n)`.

2. Find the period r of the element a such that $a^r \equiv 1 \ (mod \ N)$ via a quantum circuit. According to Shor, this takes `O((log n)`^{2}`(log log n)(log log log n))` steps on a quantum computer.

3. *Postprocessing*: Done in a classical computer in polynomial time `log (n)`.

But why is there so much excitement about this method? Compare its time complexity (big O) against the current classical champ: the Number Field Sieve as shown in Table 8-2 (including another fan favorite, the venerable Quadratic Sieve).

Table 8-2. *Time Complexities for Common Factorization Algorithms*

Algorithm	Time complexity
Shor's	$(\log n)^2(\log \log n) \ (\log \log \log n)$
Number Field Sieve	$\exp\left(c\left(\log n\right)^{1/3} \left(\log\log n\right)^{\frac{2}{3}} \right)$
Quadratic Sieve	$\exp\left(\sqrt{\ln n \ln \ln n} \right)$

Incredibly, Shor's algorithm has a polynomial time complexity, far superior to the exponential time by the Number Field Sieve, the fastest known method for factorization in a classical computer. As a matter of fact, experts have estimated that Shor's could factor a 200+ digit integer in a matter of minutes. Such a feat would rock the foundation of current asymmetric cryptography used to generate the encryption keys for all of our web communications.

Tip Symmetric cryptography is highly resistant to quantum computation and thus to Shor's algorithm.

But don't panic yet; a practical implementation in a real quantum computer is still a long way. Nevertheless, the algorithm can be simulated in a classical system using the slick Python library: ProjectQ. We'll run ProjectQ's implementation in a further section, but next let's see how period finding can solve the factorization problem efficiently.

Period Finding

Period finding is the basic building block of Shor's algorithm. By using modular arithmetic, the problem is reduced to finding the period (r) of the function $f(x) = a^x \mod N$ (see Figure 8-9).

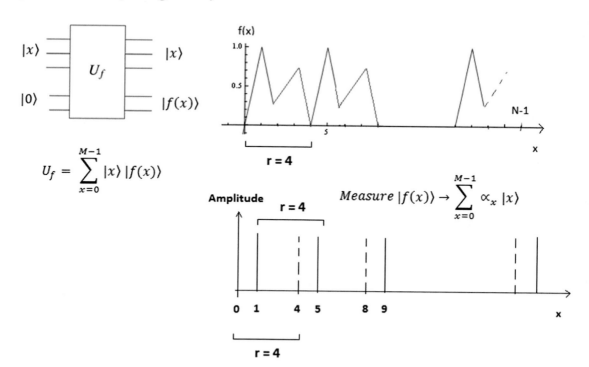

Figure 8-9. *Periodic function f(x)*

Figure 8-9 gives an example of a periodic function f(x) with period r = 4. For the algorithm to work, f(x) must meet three conditions:

1. f(x) is one-to-one on each period; that is, the values of f(x) must not repeat. In Figure 8-9 these values are represented by the vertices of each line per period.

2. For any given M or the number of periods, r must divide M. For example, given M = 100 and the period r = 4, M/r = 25.

3. M divided by r must be greater than r. That is, $M > r^2$.

Shor's algorithm transforms f(x) into a quantum circuit U_f where the inputs are in superposition. If we measure the second register in U_f, we may see values for the amplitudes $\sum_{x=0}^{M-1} \propto_x |x\rangle$ as shown in the amplitude chart of Figure 8-9. Here the amplitudes are exactly four units apart which is the period we are looking for. In this particular case, we get periodic superpositions with r = 4. But what do we do with this periodic superposition? Shor's relies on another trick: Fourier sampling or quantum Fourier transform.

Fourier Sampling

Fourier sampling is a data manipulation process that has the following properties:

- It allows for input shifting without changing the output distribution.

- This is good because now we have a periodic superposition where the non-zero amplitudes are the multiples of the period (see Figure 8-10).

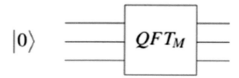

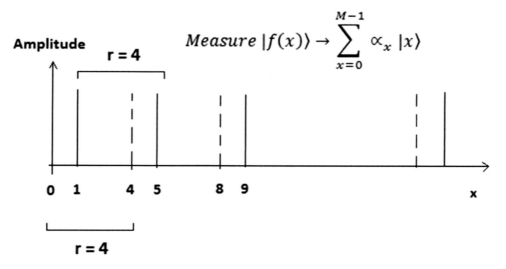

Figure 8-10. *Fourier sampling showing periodic superposition*

But what is the output of Fourier sampling? And how does it help? The answer is that its output is a random multiple of M/r. In this case given M = 100 and r = 4, we get a random multiple of 100/4 = 25. This is advantageous for our goal. Let's see how.

Feed the Fourier Sampling Results to Euclid's Greatest Common Divisor

If we were to run Fourier sampling multiple times, we will get random multiples of M/r. For example, we may get 50, 75, 25, etc. Now, if we apply Euclid's greatest common divisor (gcd) to our random outputs, then viola: By dividing M by the gcd, we get the period r. Thus

$$r = M/\gcd(50, 75, \ldots) = 100 /25 = 4$$

So this is the outline for period finding via a quantum circuit. To understand how this method can find a factor efficiently, let's run through an example by factoring the number N = 21. Our task relies on two very efficient operations:

- Modular arithmetic: a = b (mod N). For example, 3 = 15 (mod 12).

- Greatest common divisor gcd(a, b). For example, gcd(15, 21) = 3.

Thus for N = 21, we need to solve the equation $x^2 \equiv 1 \ (mod \ 21)$. That is, find the nontrivial square root x such that

- N divides (x +1) (x − 1).

- N does not divide (x ± 1).

- Finally, recover a prime factor by applying gcd(N, x+1).

To find the nontrivial factor for N = 21, pick a random x. For example, given N = 21, choose x = 2; thus

```
2⁰ ≡ 1  (mod 21)
2¹ ≡ 2  (mod 21)
2² ≡ 4  (mod 21)
2³ ≡ 8  (mod 21)
2⁴ ≡ 16 (mod 21)
2⁵ ≡ 11 (mod 21)
2⁶ ≡ 1  (mod 21). Got the period r = 6.
```

In this case, $2^6 = (2^3)^2$. Thus $2^3 = 8$ is a nontrivial factor such that 21 divides (8 + 1) (8 − 1). Finally we recover a factor with the greatest common divisor gcd (N, x+1) = gcd(21, 9) = 3. In general terms, pick an x at random, and then loop through $x^0, x^1,..., x^r \equiv$ mod N. If we are lucky, then r is even, that is, $(x^{r/2})^2 \equiv 1 \ (mod \ N)$. And thus we have a nontrivial square root of 1 mod N.

Tip   It has been proven that the probability that we get lucky, that is, r is even for $x^2 \equiv 1 \ (mod \ N)$, is ½. If we are unlucky, on the other hand, then we must repeat the procedure all over again. However given the high probability of success, this would be insignificant in the great scheme of things.

Now, let's run the algorithm using the slick Python library ProjectQ.

Shor's Algorithm by ProjectQ

ProjectQ is an open source platform for quantum computing that implements Shor's algorithm using the circuit proposed by Stéphane Beauregard[2]. This circuit uses 2n + 3 qubits where n is the number of bits of the number N to factor. Beauregard's method is divided into the following steps:

1. If N is even, return the factor 2.

2. Classically determine if **N = p^q for p ≥ 1 and q ≥ 2**, and if so, return the factor p (in a classical computer, this can be done in polynomial time).

3. Choose a random number a, such that **1 < a ≤ N – 1**. Using Euclid's greatest common divisor, determine if **gcd (a, N) > 1**. If so, return the factor **gcd(a, N)**.

4. Use the order-finding quantum circuit to find the order r of **a modulo N**. In a quantum computer, this step is done in polynomial time.

5. If r is odd or r is even but **$a^{r/2}$ = –1 (mod N)**, then go to step 3. Otherwise compute **gcd($a^{r/2}$ – 1, N)** and **gcd($a^{r/2}$ + 1, N)**. Test to see if one of these is a nontrivial factor of N, and return the factor if so (in a classical computer, this can be done in polynomial time).

---

[2]Stéphane Beauregard, Circuit for Shor's algorithm using 2n+3 qubits. Département de Physique et, Université de Montréal.

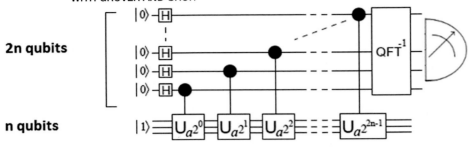

Controlled Multiplier

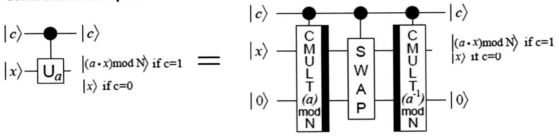

Figure 8-11. *Beauregard quantum circuit for period finding*

Beauregard implements period finding by using a series of controlled additions and multiplications in Fourier space to solve $f(x) = a^x(mod\ N) \rightarrow a^r \equiv 1\ mod\ N$ (see Figure 8-11):

- A controlled multiplier U_a maps $|x\rangle \rightarrow |\ ax\ (mod\ N)\rangle$ where

    - a is a classical relative prime to use as the base for a^x (mod N).

    - x is the quantum register.

    - c is the register of control qubits such that $U_a = ax$ (mod N) if c =1 and x otherwise.

- The controller multiplier U_a is in turn implemented as a series of doubly controlled modular adder gates such that

    - If both control qubits c1 = c2 = 1, the output is $f(x) = |\ \varphi(a+b\ mod\ N)\rangle$. That is, a + b (mod N) in Fourier space. Note that this gate adds two numbers: a relative prime (a) and a quantum number (b).

    - If either control qubit (c1, c2) is in state |0>, then $f(x) = |\ \varphi(b\ )\rangle$.

- The doubly controlled modular adder gate is in turn built on top of the quantum addition circuit by Draper[3]. This circuit implements addition of a classical value (a) to the quantum value (b) in Fourier space.

Factorization with ProjectQ

Let's install ProjectQ and put the algorithm to the test. The first thing to do is to use the Python package manager to download and install ProjectQ (note that I am using Windows for the sake of simplicity. Linux users should be able to follow the same procedure):

```
C:\> pip install projectq
```

Next, grab the shor.py script from ProjectQ's examples folder[4] or the book source under Workspace\Ch08\p08-shor.py. Now, run the script and enter a number to factor (see Listing 8-2).

Listing 8-2. Shor's Algorithm by ProjectQ in Action

```
C:\>python shor.py
Number to factor: 21

Factoring N = 21: .........

Factors found : 7 * 3 = 21
Gate class counts:
    AllocateQubitGate : 166
    CCR : 1467
    CR : 7180
    CSwapGate : 50
    CXGate : 200
    DeallocateQubitGate : 166
    HGate : 2600
```

---

[3]T. Draper (2000), Addition on a quantum computer, quant-ph/0008033. Available online at https://arxiv.org/abs/quant-ph/0008033.

[4]ProjectQ – an open source software framework for quantum computing. Available online at https://github.com/ProjectQ-Framework/ProjectQ.

```
    MeasureGate : 11
    R : 608
    XGate : 206

Gate counts:
    Allocate : 166
    CCR(0.098174770425) : 18
    CCR(0.196349540849) : 30
    CCR(0.392699081699) : 70
    CCR(0.490873852124) : 18
    CCR(0.785398163397) : 80
    CCR(0.981747704246) : 38
    CCR(1.079922474671) : 20
    CCR(1.178097245096) : 16

      ...
    R(5.252350217719) : 1
    R(5.301437602932) : 1
    R(5.497787143782) : 1
    X : 206

Max. width (number of qubits) : 13.
--- 5.834410190582275 seconds ---
```

For N = 21, the script dumps a set of very helpful statistics such as

- *The number of qubits used*: Given N = 21 we need 5 bits; thus total-qubits = 2 * 5 + 3 = 13.

- *The total number of gates used by type*: In this case, doubly controlled CCR = 1467, CR = 7180, CSwap = 50, CX = 200, R = 608, X = 206, and others, for a grand total of 12,646 quantum gates.

ProjectQ implements quantum period finding using Beauregard algorithm as shown in Listing 8-3:

- The run_shor function takes three arguments:

    - The quantum engine or simulator provided by ProjectQ plus

    - N: The number to factor

    - a: The relative prime to use as a base for a^x mod N

- The function then loops from a = 0 to a = ln(N) with the quantum
  input register x in superposition; it then performs the quantum
  circuit for f(a) = a^x mod N as shown in Figure 8-11.

- Next, it performs Fourier sampling on the x register conditioned on
  previous outcomes and performs measurements.

- Finally it sums the measured values into a number in range [0,1]. It
  then uses continued fraction expansion to return the denominator or
  potential period (r).

Listing 8-3. ProjectQ Period Finding Quantum Subroutine

```
def run_shor(eng, N, a):
  n = int(math.ceil(math.log(N, 2)))

  x = eng.allocate_qureg(n)

  X | x[0]

  measurements = [0] * (2 * n)  # will hold the 2n measurement results

  ctrl_qubit = eng.allocate_qubit()

  for k in range(2 * n):
    current_a = pow(a, 1 << (2 * n - 1 - k), N)

      # one iteration of 1-qubit QPE
    H | ctrl_qubit

    with Control(eng, ctrl_qubit):
      MultiplyByConstantModN(current_a, N) | x

    # perform inverse QFT --> Rotations conditioned on previous outcomes
    for i in range(k):
      if measurements[i]:
        R(-math.pi/(1 << (k - i))) | ctrl_qubit

    H | ctrl_qubit
```

```
    # and measure
    Measure | ctrl_qubit
    eng.flush()
    measurements[k] = int(ctrl_qubit)
    if measurements[k]:
      X | ctrl_qubit

  Measure | x
  # turn the measured values into a number in [0,1)
  y = sum([(measurements[2 * n - 1 - i]*1. / (1 << (i + 1)))
       for i in range(2 * n)])

  # continued fraction expansion to get denominator (the period?)
  r = Fraction(y).limit_denominator(N-1).denominator

  # return the (potential) period
  return r
```

The next section compiles a set of factorization results for various values of N.

Simulation Results

ProjectQ's period finding subroutine is a simulation of a quantum circuit on a classical computer so it is not practical to use it to factorize large numbers. As a matter of fact, it is not capable to factor numbers larger than four digits in reasonable time on a home PC. Table 8-3 shows a set of results for various values of N gathered from my laptop up to 2491.

Table 8-3. *Factorization Results for Various Values of N*

Number (N)	Qubits	Time (s)	Memory (MB)	Quantum gate counts
15	11	2.44	50	CCR = 792 CR = 3186 CSwap = 32 CX = 128 H = 1408 R = 320 X = 130 Measure = 9
105	17	27.74	200	CCR = 3735 CR = 25062 CSwap = 98 CX = 392 H = 6666 R = 1568 X = 393 Measure = 15
1150	25	17542.12 (4.8 h)	500	CCR = 15366 CR = 139382 CSwap = 242 CX = 968 H = 24222 R = 5829 X = 981 Measure = 23
2491	27	246164.74 (68.3 h)	2048	CCR = 20601 CR = 194670 CSwap = 288 CX = 1152 H = 31126 R = 7509 X = 1166 Measure = 25

Factorizing the four-digit number 2491 took more than 68 hours on a 64-bit Windows 7 PC with an Intel Core i-5 CPU at 2.6 GHz with 16 GB of RAM. I tried to go a bit higher by attempting to factorize N = 8122 but gave up after one week. All in all, these results show that the algorithm can be simulated successfully for small numbers of N; however it needs to be implemented in a real quantum computer to test its real power.

This chapter brought proceedings to a close with two algorithms that have generated excitement about the possibilities of practical quantum computation: Grover's algorithm, an unstructured quantum search method capable of finding inputs at an average of square root of N steps. This is much faster than the best classical solution at an average of N/2 steps. Expect all web searches to be performed by Grover's algorithm in the future.

Shor's algorithm for factorization in a quantum computer which experts say could bring current asymmetric cryptography to its knees. Shor's, arguably the most famous quantum algorithm out there, is a prime example of the power of quantum computation by providing exponential speedups over the best classical solution.

Finally, I would like to close things up by saying that I have tried my best to explain the difficult concepts of quantum computing by mixing math, software, and as many figures I can muster. A lot of coffee cups and sleepless nights were spent writing this manuscript, not to mention that I find most of the math as confusing as you probably do. I hope you enjoy reading this book as much as I did writing it, and remember: The great physicist Richard Feynman once said "If somebody tells you that he understands Quantum Mechanics it means he doesn't understand Quantum Mechanics."

Index

A

Adiabatic quantum
     computation (AQC), 68
Adiabatic theorem, 68–69
Aircraft industry, 76
Alain Aspect experiment, 21
Algorithm by projectQ
    in action, 333–334
      Beauregard quantum circuit, 332–333
      factorization, 333
      simulation results, 337–338
      subroutine, 335–336
Almighty wave function, 7–8
API token, 102
Atomic theory, 8
Authentication, 102–103

B

basisGates, 88
Bell's
    inequality, 18–19
      experiment, 100
    theorem, 17
Bell states
    CHSH inequality, 91
    compiled results, 94
    correlation probability, 94
    inequality, 90
    measurements, 92

    permutations, 91
    photon polarization, 89–90
    quantum circuits, 91, 93
    qubits, 91, 92
Black-body radiation experiment, 3–4
Black holes, 62
Bloch sphere, 40
Bohr/Heisenberg interpretation, 13
Bohr's
    atom, 10
    quantum jump, 4
Boltzmann's entropy, 4
Born's probabilistic party, 9–10
Bosonic codes, 66
Boson sampling problem, 60

C

Calibration parameters, 106–108
Catastrophic paradox, 2
CentOS, 145, 264
cgi package, 266–267
Chinese academy of sciences (CAS), 76
chmod OS command, 277
Cloud quantum battleship, 267–268
    decouple, user interface, 237–238
    features, 223
    HTTP request, 243–244
    J2EE project, 276
    JSON document, 247

© Vladimir Silva 2018
V. Silva, *Practical Quantum Computing for Developers*, https://doi.org/10.1007/978-1-4842-4218-6

S

T

U, V

46115153R00203

Made in the USA
San Bernardino, CA
03 August 2019